# PRAISE FOR *THE GOOD VIRUS*

'One of the best books of *any* genre that I've read in 2023, this superbly-written book relies on exquisite story-telling to interweave science and history and politics into an engaging and readable account that will fascinate absolutely *everyone*. Whether you are looking for something unique to enthrall your book club friends, something educational to enlighten or inspire . . . or insights into the complex and subtle ways that politics, history, medicine, science and individual personalities all feedback on and influence each other, you will find it in this remarkable and extraordinarily readable book. Even scientists and medical doctors will find much in this book to intrigue and delight them, and non-specialists will find this eye-opening book is unlike anything they've ever read before.'

*Forbes*

'Most viruses do no harm to humans – and, as this fascinating book explains, a large class of them might even prove our saviours . . . [Phages] regulate our gut microbiome, are crucial to marine ecosystems, and inspired the modern Crispr technique of gene-editing . . . All this and more is thrillingly recounted in Tom Ireland's superb book. This is real luxury-class science writing, exploring how a "Stalin-tainted" idea from long ago can be rehabilitated, alternating scientist interviews and vivid case studies of miraculous-seeming cures with historical narrative and limpid biotechnological explanations . . . He also demonstrates excellent comic timing.'

Steven Poole, *Telegraph*

'This engaging book highlights the brighter side of the viral world . . . a delight. To learn more about phages is to discover fascinating details about a hidden world . . . Ireland offers riveting accounts . . . *The Good Virus* is timely . . . It's an exciting time for a field that has, for too long, been unfairly overlooked.'

*Nature*

'Tom Ireland's compelling and original book makes a strong case for revisiting phage therapy . . . richly detailed and absorbing, and well balanced between the biological details and the personalities and scientific politics involved . . . *The Good Virus* is original, eye-opening and grippingly told.'

*New Scientist*

'*The Good Virus* is a colorful redemption story for the oft-neglected yet incredibly abundant phage, and its potential for quelling the existential threat of antibiotic resistance . . . Ireland, an award-winning science journalist, approaches the subject of his first book with curiosity and passion, delivering a deft narrative that is rich and approachable . . . Ireland tells the fascinating story of how phages harvested from German corpses helped the Soviets defeat the Nazis when cholera broke out during the siege of Stalingrad.'

*New York Times*

'In the wake of the Covid pandemic, the idea of a virus being beneficial may seem strange, even implausible. But science journalist Tom Ireland is admirably determined to show us just how potent this disease-fighting approach can be and to persuade us of its importance. As engaging as it is expansive, *The Good Virus* describes the distinctive biology and murky history of bacteriophage (generally shortened to "phage"), a form of life that is remarkably abundant yet obscure enough to have been termed the "dark matter of biology."'

*Wall Street Journal*

# The Good Virus

*The Mysterious Microbes that
Rule Our World, Shape Our Health
and Can Save Our Future*

Tom Ireland

hodder
press

First published in Great Britain in 2023 by Hodder & Stoughton
An Hachette UK company

1

Copyright © Tom Ireland 2023

All illustrations (in Field Guide to Phages) copyright © Tom Ireland 2023

With thanks to Yale University Press for permission for all quotations
from *W. C. Félix d'Herelle and the Origins of Molecular Biology* (New Haven
and London: Yale University Press, 1999) by William C Summers.
Reproduced with permission of the Licensor through PLSclear.

A CIP catalogue record for this title is available from the British Library

Paperback ISBN 978 1 529 36528 3
eBook ISBN 978 1 529 36526 9

Typeset in Bembo MT by Hewer Text UK Ltd, Edinburgh
Printed and bound in Great Britain by Clays Ltd, Elcograf S.p.A.

Hodder & Stoughton policy is to use papers that are natural, renewable
and recyclable products and made from wood grown in sustainable forests.
The logging and manufacturing processes are expected to conform
to the environmental regulations of the country of origin.

Hodder & Stoughton Ltd
Carmelite House
50 Victoria Embankment
London EC4Y 0DZ

www.hodder.co.uk

*For Melon*

# Contents

## Part 4: Fundamental Phages

## Part 5: Future Phages

# Author's note

Scattered throughout this book are dramatic examples of people receiving or seeking experimental medical treatments with viruses. These are not intended to represent, nor should they be read as, a body of evidence for the efficacy of these treatments. They are included to help demonstrate what patients and doctors must resort to when our existing antibiotics fail, and why the development of a modern, safe and clinically proven version of these strange and controversial treatments is so desperately needed.

# Introduction

## Invisible allies

In the summer of 1942, as German troops encircled the Russian city of Stalingrad, Nazi commanders began to receive bizarre reports of dead bodies disappearing from German field hospitals. In the dead of night, Soviet scouts were crossing the front line daringly to steal certain German corpses, before squirrelling them back down into a secret underground laboratory hidden deep beneath the city.

The Germans had been suffering outbreaks of cholera for weeks as they advanced east towards Stalingrad, and the Soviets were desperate to prevent the disease from crossing the front line. Although this nasty bacterial disease had helped further deplete their enemy, it could also spread like wildfire among the soldiers and civilians crammed into a city under siege. So what on earth were the Soviets doing seeking out potentially infectious German corpses and dragging them into their territory?

Even with today's improved sanitation and modern antibiotics, cholera kills over 100,000 people every year. Spread through water contaminated with the tiny tampon-shaped bacteria *Vibrio cholerae*, if left untreated it causes debilitating cramps, diarrhoea, dehydration and, eventually, shock, coma and death. Professor Zinaida Yermolyeva, from Moscow's

Institute of Experimental Medicine, had been sent by Joseph Stalin to assess the cholera outbreaks on the front and formulate a plan.

Like other physician-scientists of the era, she had spent her career in a world without effective antibiotics, trying to work out how to kill bacteria like *Vibrio cholerae* without also killing the people infected with it. In 1942, just one genuine antibiotic substance was known to scientists – penicillin – and they were still unable to produce it in large enough quantities to treat patients. Most treatments for bacterial disease at the time were inconsistent, toxic, useless or all three. But one way to treat bacterial disease had shown more promise, especially in war, and Yermolyeva had become an expert in deploying it under battlefield conditions. It required cultivating a natural and yet invisible enemy of the cholera bacteria, which she could only find on the bodies of those who had cholera, or were close to those who had it. And so her morbid plan began. She was going to use *viruses* to kill the bacteria that were killing soldiers.

Most people have quite a low opinion of viruses. It's understandable – viruses make us ill, disable us or kill us; they spoil our crops and kill our animals. Computer viruses mess up our expensive machines and send emails that make us look like idiots. In 1985, the biologist Sir Peter Medawar famously described a virus as 'a piece of bad news wrapped in protein'. The COVID-19 pandemic has further reinforced the idea that these tiny life forms exist only to spread pestilence, ill-health, economic disaster and death. Perhaps now is as good a time as ever to explain why that view is wrong.

I first considered writing a book on 'good viruses' in early 2020. The idea was to celebrate one particularly important but overlooked group of viruses with an intriguing and controversial history, which, fascinatingly, were once used to save lives – but then were largely forgotten. At that point, where I live in the UK, the public health messaging on the novel coronavirus spreading around the world from Wuhan, China, was essentially that if we all just wash our hands more often, then it should all be fine. Like most people, I had no idea how this one extremely not-good virus would come to dominate so many lives, for so long. And so, I soon found myself writing a book celebrating viruses as the worst viral pandemic in living memory swept around the world. Indeed, much of this book was written while my family and I hid from a virus.

I still believe viruses can be, and are, 'good'. Thus the title of this book is not meant to be provocative or to diminish the immense suffering that viruses have caused – and continue to cause – us and our loved ones. Viruses kill millions of people every year and cause poor health in even more, whether it be in acute and terrifying infections like Ebola, familiar and ongoing global threats like flu, sexually transmitted viruses like herpes, childhood diseases like mumps, the millions of cases of sickness and diarrhoea caused by viruses like norovirus or dysentery or entirely new pathogens like COVID-19. Our battle with viruses has shaped human history, often tragically. However, the truth is that the viruses that cause us such suffering are vastly outnumbered by viruses that do extraordinary things for us: there are trillions of viruses out there that could actually save lives.

There are several types of virus that could be described as 'good viruses' – the harmless viruses used in many life-saving vaccines, for example, and the viruses scientists have developed that infect and kill only tumour cells, known as oncolytic viruses – a promising new therapy for cancer. There are the ancient viruses that became embedded in the genomes of our ancestors, which make us the species we are today. Several crucial genes that all mammals have are derived from viral genes that at some point were assimilated into our ancestors' DNA. For example, genes that help the cells of a foetus connect to its mother's cells in the placenta are derived from a gene that helps viruses fuse with cells. (Without this gene, we might still be laying eggs.) Another, essential for long-lasting information storage in the mammalian brain, is believed to be a repurposed gene that evolved in a virus to help it encapsulate genetic material.

While these old and new viruses and their various functions are fascinating, in this book, I focus on a very special group of viruses known as bacteriophages. These were the viruses Professor Yermolyeva was looking for, some eighty years ago, in the icy soil around Stalingrad.

Bacteriophages, known simply as 'phages' for short, are viruses that infect and kill bacteria. Essentially harmless to humans, they exist solely to inject their genes into bacterial cells, where they can either lurk indefinitely or replicate madly. In the latter case, they cause their unfortunate host's metabolism to go haywire, churning out copies of the virus instead of the materials it needs to sustain itself. When the time is right, the new viruses burst the bacteria open like a popped water bomb and spill out to find new hosts in which to repeat the process.

The vast majority of these viruses are so-called 'tailed' phages. Along with a sinister-looking twenty-faced head, known as a capsid, these viruses have a distinctive protein tube, or tail, which they use to inject their DNA into their unfortunate host like a tiny syringe. Even finer, spider-like legs fold out from the base of the phage to help it detect and bind to the surface of the bacterial cell, like an unfathomably tiny lunar lander.*

Confusingly, the two major types of microbes at war in this book – bacteria and viruses – are often grouped together simply as 'germs', but they are distinct in important ways. The most basic difference between them is that bacteria are cells and viruses are not. Cells are biology's basic units of life – microscopic capsules with everything needed for life and replication contained within a fat-based membrane and, sometimes, a tough outer wall. All life on the planet – except viruses – consists of cells, either working in concert with one another (like the human body, for example, a network of trillions of related cells arranged to form tissues and organs) or existing just fine as single cells.

Viruses, conversely, are far less complex. At their simplest, they are little more than a length of genetic material (normally DNA, deoxyribonucleic acid, but sometimes its chemical cousin, RNA, ribonucleic acid) wrapped in a protective protein capsule. Outside of a host, they are inert, lifeless even, lacking the biochemical components to do anything with the information contained in their genes.

In order to replicate, the virus must get inside a cell. Viruses have been described as living 'a kind of borrowed life', only

---

* Some scientists in the phage community have suggested that the design of the 1968 Apollo lunar module was actually inspired by phages – the lander's squat shape being so similar to stubby tailed phages known as *podoviruses*.

ever able to exert an influence on the world when inside a host cell. It's a little like how a computer virus is just a piece of code on a USB stick – unable to do anything when lying in a drawer – until it is placed into a computer, when it can suddenly instruct that computer's systems to send copies of itself to a thousand inboxes around the world. This reliance on other life is, in part, why there has always been a debate over whether viruses are 'alive' or not. To me, the question is unhelpful, suggesting that viruses are somehow not a proper, paid-up member of our wonderful living world.

Whether or not viruses meet the criteria we have decided characterises a distinct living being, they are an essential biological component of the ecosystems that have developed on Earth. They are built from the same basic building blocks as life, use the same chemical language as life, evolve and replicate alongside life, and interact with and transform life. Some scientists believe that all life may have evolved from self-replicating entities more akin to viruses than cells.[1] And by operating in the fascinating and illuminating grey area where complex chemistry becomes simple biology, they can arguably tell us more about what life is than living creatures so complex that they may never be fully understood.

For every type of cellular life on earth – bacterial, fungal, animal or plant, and the weird things somewhere in between – there are viruses that have evolved to infect them, and together these viruses outnumber all other living entities on Earth. While we commonly associate viruses with disease and death, just a tiny fraction are a danger to us. The vast majority are phages. And it is only very recently that we have begun to understand that phages are an essential part of the living fabric of the planet, drivers of innovation, diversification and change.

Bacteria are also essential to all life on Earth. Although we have 'learnt to hate and fear them', as the science writer Ed Yong puts it,[2] just a hundred or so of the many thousands of species of bacteria in the world colonise our body in a way that makes us ill or causes disease. Even these mostly live quite happily on and around us without our noticing, only causing ill health when a vulnerability in our immune systems is exposed. The rest perform a suite of essential environmental services that make our planet hospitable. They capture chemical and solar energy to form the foundational layer of the food chains that support the rest of life on Earth; they take inorganic material, other life's waste-products and dead things and recycle them back into forms that can be used by other life. They produce 20% of the atmospheric oxygen we breathe. They help us digest our food, help plants absorb nutrients, protect us from other microbes and ferment some of our favourite foods. They have adapted and co-evolved to live in almost every environmental niche on the planet, from boiling vents at the bottom of the sea to the internal organs and tissues of other life, from lakes with the acidity of battery acid to the nodules in the roots of our most important crops.

They, and other similar single-celled life that together are known as prokaryotes, have been growing and multiplying on Earth ceaselessly for almost four billion years, since life first emerged on our scorching, primordial rock. Among the most ancient forms of life on the planet, they have evolved into thousands, probably millions of different species, exploiting and colonising virtually every environment possible. They are literally everywhere. Just on the sponge in your kitchen sink, there are probably more bacteria than the total number of humans who have ever lived.

For as long as all this bacteria has been around, however, phages have been perfecting the art of infecting and destroying them. For every single one of the immense number of bacterial cells on the planet, there are thought to be at least ten phages – perhaps more. Anywhere and everywhere a bacterial strain has evolved to exploit an ecological niche, there will be viruses that have evolved to exploit that bacteria. This makes these seemingly obscure viruses easily the most abundant biological entity on Earth.

Sail out into the middle of the ocean and scoop up a cup of water and it will contain millions, possibly hundreds of millions of phages. Take some water from a briny marsh or your local stream, a caustic alkaline lake or a scorching hydrothermal vent and still you'll find millions of phages in every millilitre. On land, there can be even higher concentrations – billions of phages in a single gram of rich soil. Even a gram of baked desert earth or frozen Arctic peat contains an active community of millions of phages, locked in a never-ending dance with their bacterial hosts.

There are so many phages on this planet that they can even be found floating in thin air. When one group of researchers installed collection devices on a concrete platform almost 3km above sea level in the Sierra Nevada mountains in Spain, they found that hundreds of millions and sometimes billions of viruses[3] rained down onto their equipment every day.* Researchers estimate there may be as many as $10^{31}$ phages on Earth – that's 10 with 30 zeros after it – a truly preposterous number that equates to around a trillion phages for every grain

---

* Probably blown high into the air from ocean spray or high winds blowing phages off topsoil and crops.

of sand on the planet. The biologist J. B. S. Haldane famously quipped that if God created all the living organisms on Earth, then the creator must have 'an inordinate fondness for beetles': it seems God is even fonder of bacterial viruses.

Of course, these viruses are all over and inside our bodies, doing their deadly bacteria-bursting thing millions of times over as you read this introduction. It is often said that there are more bacterial cells in a human body than human cells. Well, there are even more phages: trillions and trillions of them in our guts in particular. These phages are bursting bacteria open inside you and all around you right now, and in every moment of your life.

And so, when Professor Zinaida Yermolyeva decided, all those years ago, to try to use viruses to kill the bacteria threatening to wipe out the soldiers defending Stalingrad, the problem wasn't finding one – it was finding the right one.

Today, in a few scattered pockets of the former Soviet Union, swigging a tiny vial of yellowish liquid, thick with trillions of phages, is just as common as taking an antibiotic pill. In parts of Georgia and Russia, you can buy packets of phages over the counter to help treat stomach bugs, infected cuts or spots. In Tbilisi, the capital of Georgia, and in the Polish city of Wroclaw, clinics offer more intense 'phage therapy' – concentrated viruses washed straight into infected wounds or, for severe infections, injected intravenously.

Since the break-up of the USSR, in the 1990s, an increasing number of patients from all over the world, with bacterial infections that Western medicine just can't treat, have taken the long and expensive trip to these strange and often outdated clinics. Bacteria that have evolved resistance to our most important antibiotics are becoming increasingly widespread and, for a growing number of people, infections that were once easy to treat are now unstoppable.

Phages were first discovered around 1917* and were first used medically just a couple of years later, in 1919, almost twenty-five years before the first true antibiotic drug, penicillin, became widely available to doctors. For a few decades in the early twentieth century, the world went mad for phages, and phage therapy was everywhere – from chemists' shops in Britain to Brazilian public hospitals. Large European and American pharmaceutical companies mass-produced mixtures of different phages for the treatment of an assortment of bacterial diseases, and the viruses were brewed in industrial-sized copper vats by the Soviets during World War II to try to keep troops free from gangrene, cholera, dysentery and other nasty diseases of war.

But there was a catch.

You can't use just any old phage to treat any old bacterial disease. Certain phages only infect certain bacteria, and most are extraordinarily specific in which bacteria they target. Some will only infect just one or perhaps two very similar species of bacteria, and most are even choosier, infecting only a specific *strain* of a specific species. For doctors in the early twentieth century, this made phages extremely difficult to work with: if a phage they had was active against the strain of bacteria causing the disease, the results could be spectacular – patients

---

* Although the debate over who first discovered phages, and when, still rages today – as we will see later in the book.

brought back from the brink of death, up and walking and rid of their bacterial invaders completely within days. But if the phages weren't an exact match, it was completely useless.

Yermolyeva, looking for phages that could keep the people of Stalingrad safe from the cholera spreading on the German side of the front line, needed to find phages that could infect and kill the exact strain of *Vibrio cholerae* bacteria causing the local outbreaks. The best place to find these viruses was in and around the deadly bacteria itself. And the best place to find the deadly bacteria was on the bodies of people who were dying from it.

Working with the corpses in her underground lair, she soon isolated the strain of *Vibrio cholerae* bacteria causing the disease, and then the phages living alongside it. She tested which ones could kill the bacteria most effectively, and using only rudimentary equipment, isolated them, concentrated them and purified them. The phages in their natural state had clearly not helped the dead soldiers on which they were found – but Yermolyeva aimed to create a concentrated mix of the most powerful viruses that together, might be able to overpower a nascent cholera infection before it took hold. Soon she had made enough anti-cholera phage mixture to ensure 50,000 preventative doses were given out to soldiers and civilians in the city every day.[4,5]

Ahead of what would become a pivotal battle of World War II and the defeat of the German Army, Yermolyeva is said to have received a call from none other than the commander-in-chief of the Soviet Union, Joseph Stalin.* 'Is it safe to keep

* It goes without saying that accounts from within the USSR about Stalin during his reign should be taken with a pinch of salt.

more than a million people at Stalingrad? Can the cholera epidemic interfere with the military plans?' he asked. Yermolyeva replied that she was winning on her front – no cholera outbreaks had broken out within the city. Now it was the Red Army's turn to win on theirs.

It is hard for most of us to truly understand that a never-ending microbial war between viruses and bacteria is going on in the world around us, every minute of every day. Given how easily our brains are baffled by extremely large and small numbers, accurately perceiving the size of a single bacterial cell or virus in our mind's eye is tricky. Bacteria are generally measured in units known as micrometres – that is, one-thousandth of a millimetre. And viruses, far smaller, must be measured in nanometres – one-thousandth of a micrometre.

If you've no idea what that really means, practically, you're not alone. (Apparently, it's about the length that your nails grow every *second*.) To help you visualise a micrometre, the width of an average-sized human hair is around 100 micrometres. One of the largest types of human cell, a skin cell, is less than half the width of a hair's breadth – a blob just 30 micrometres across. Our eyes cannot see objects this small. These and other human cells still dwarf bacterial cells. A single rod-shaped *E.coli* bacterial cell is just three micrometres long, and less than one micrometre wide – you could fit ten *E.coli* cells lengthways inside the skin cell, and dozens across the shaft of a human hair.

To understand the size of a virus, scientists have to go down to the nanoscale – a set of units used to measure the distance between atoms. A nanometre is a *thousandth of a micrometre*. The T4 phage, a well-studied virus of *E.coli* bacteria, is a fairly large virus at 200 nanometres long and 70 nanometres wide. The *E.coli* cell, at 3,000 nanometres long, now looks like a giant hairy bean in comparison.

The extremely tiny size of viruses – smaller than the wavelength of light itself – means that they remained mysterious to science for decades into the twentieth century, hundreds of years after we were able to see bacteria with microscopes. Modern microscopes that use beams of electrons, rather than light, now allow us to 'see' these tiny predators, stuck into their bacterial host cells like pins in a voodoo doll.

With the help of this technology, we see that phages, although tiny and relatively simple compared to other forms of replicating life on Earth, are exquisitely, beautifully evolved. They come in a variety of shapes and sizes, with whiskers, tails, spider-like legs and landing gear for attaching and breaking into their hosts.

These tiny protein structures help the all-important phage genes survive in the world and enter a suitable host. Once the phage chemically binds itself to its host, its genes are pumped into the cell under huge pressure, along with a barrage of chemicals to disrupt the bacterium's defences. As the hijacked cell starts synthesising the component parts of new viruses, those parts then self-assemble into new phages, a mind-blowing feat of bioengineering that scientists don't yet fully understand. Finally, in the most spectacular act of the infection cycle, the phage genes instruct the host to synthesise a suite of enzymes to rupture itself from the inside out, a violent end meticulously timed to maximise the number of viruses released.

Once we see the viruses all around us in this way – as molecular machines, capable of manipulating cells and the building blocks of life in an impossibly intricate and sophisticated way – how can we not be filled with wonder?

Phages are one of the most important and yet underappreciated life forms on this planet. The totality of what these viruses do and *could* do for us is immense, but remains largely unknown and uncelebrated. These tiny biological predators keep bacterial growth in check in every known ecosystem, and the endless release of new phages from infected bacteria creates the extraordinary numbers of phages on this planet. If you could lump all these tiny slivers of protein and DNA together, they would weigh the same as roughly 75 million blue whales[6], or for those of us who use London Buses as their imaginary units, over 800 million Routemasters.

Constantly hijacking and breaking open the microorganisms that cycle nutrients and energy in the environment, phages are thought to be responsible for approximately 10% of the carbon turnover of the entire planet. By constantly swapping genes with their hosts and each other, phages have played a central role in the spread of important genetic traits around the world. And, because bacteria must constantly change and evolve to evade them, they have helped drive the diversity and complexity of life on Earth that led to higher organisms, including us.

You may think that, considering that we're covered in billions of these nanoscopic killing machines every day of our lives, we should know quite a lot about them. But here's the wild thing: we know hardly anything about any of them. Together, phages represent the greatest source of genetic diversity on the planet, and it is estimated that something like 99% of them may be unknown to science.

While scientists have been studying microorganisms like bacteria since the late 1600s, it is only in the last thirty years that they have realised that our seas and soils and bodies are awash with bacterial viruses. Even the phages that call our guts home, which help regulate the microbial communities now known to be crucial to our health and wellbeing, are mostly unknown to science. Ask most people what a phage is and they will look at you blankly. It's likely that more people have heard of the Marvel Comics character Phage than actual phages, the most abundant and diverse form of life on the planet.

The more we study phages, the more remarkable things we discover. CRISPR, the revolutionary gene-editing technology that in the last decade has dramatically changed what is possible in biotechnology and medicine, uses at its heart a microbial immune system that evolved to defend cells from attack by phages. Up to an eighth of the oxygen in the atmosphere is thought to be produced thanks to genes that marine phages give to their bacterial hosts. And so-called jumbo phages – relatively massive phages which seem to have some of the complexity of cellular life – could even help us answer one of the great mysteries of how the first cells on Earth originated. We are only just beginning to understand how the soup of viruses in which we all live impacts us and our planet and how we can use their power for good.

Perhaps as remarkable as phages themselves is the strange and often dramatic history of our relationship with them. As we saw earlier, phages were first deployed in medicine almost one hundred years ago, were once used to treat infectious disease all across the globe, and arguably helped turn the tide of World War II. So why was this important idea abandoned? How could it be virtually forgotten by the entirety of western medicine? And why is our understanding of these ubiquitous, important viruses still in its infancy?

The answer is not just about science. This book is about how personalities, power, politics – and our narrow view of viruses – have repeatedly crashed together to hinder our understanding of what phages are and how we could benefit from their bacteria-killing abilities. This is an epic story with a cast of brilliantly bold and often eccentric scientists. While most focused their attention on the viruses that harm humans, these extraordinary characters focused on the seemingly obscure viruses of bacteria, often in the face of indifference and even hostility to their work.

The story starts with a decades-long effort by some of the world's leading microbiologists to deny that these viruses even existed. We'll explore why phage therapy flourished in Stalin's Soviet Union, before the regime went murderously mad and almost destroyed its world-beating phage expertise. We'll examine why the rest of the world shunned the idea of using phages in medicine for much of the last century, and the heroes and outsiders and who fought to keep the idea alive.

We'll go phage-hunting with the surprisingly small community of scientists now tasked with finding and cataloguing the dizzying diversity of phages on our planet, from the frozen poles to the bottom of the sea, and we'll learn how phages have been central to some of the most important and exciting scientific breakthroughs of the past hundred years. Along the way, we'll meet a range of very special phages, from some of the simplest replicating life forms on Earth to the 'giant phages' that blur the line between viral and cellular life, as well as other less well-known but equally exceptional bacterial viruses, each with their own unique strategies for exploiting their bacterial hosts.

And, most crucially, we will see why we need phages more than ever before: the escalating threats posed by antibiotic-resistant bacteria.

In February 2022, as many countries began to finally see a path out of the COVID-19 pandemic, a report by the medical journal *The Lancet* revealed the terrible scale of a parallel, silent and perhaps more worrying public health problem, a pandemic involving not one pathogen, but many. The report pooled data from across the world to try to understand, for the first time, how many people were dying each year from bacterial infections that were no longer treatable with antibiotic drugs. They found that in 2019 alone, almost two million people had died from infections with bacteria that were resistant to frontline antibiotics, and a further three million died from causes associated with a drug-resistant infection.[7] That puts drug-resistant infections up there with killers like diarrhoeal diseases and dementia on the World Health Organization's (WHO) list of the ten leading causes of death globally.[8]

Antibiotics revolutionised healthcare in the mid-twentieth century, but they are not the panacea we once thought they would be. Some bacteria are naturally resistant to or unaffected by certain antibiotics, while others are able to surround themselves in a thick, sticky slime as they colonise surfaces in our bodies, forming what is known as a biofilm. This not only helps firmly adhere the growing colony in place but also creates a tough physical barrier to keep antibiotics out. Antibiotics also destroy many of the bacteria that we need to keep us healthy, causing both temporary and sometimes long-term side-effects – the medical equivalent of a bomb dropped on a city to weed out one group of troublemakers.

More importantly, we have overused these special compounds, in both medicine and agriculture, and allowed them to wash out into the world in industrial quantities. This has created environments – both in hospitals and in our waterways – where any bacteria with some degree of tolerance to these chemicals becomes dominant, and stronger and stronger resistance to the drugs evolves. Bacteria live and die fast, and therefore evolve fast; they have an annoying habit of trading useful genes between themselves, meaning resistance to antibiotics emerges and spreads quickly and repeatedly.

For much of the late twentieth century, as resistant strains of bacteria emerged and spread, scientists simply developed new types of antibiotics to keep bacteria guessing. But that pipe has run dry – no major new class of antibiotic has been developed in the past thirty years. In recent decades, even our 'last-resort' antibiotics, reserved for mortal emergencies, have alarmingly started to fail. The continued overuse of antibiotics in medicine and in agriculture is now causing a boom in the number of bacterial strains with resistance to not just a few common drugs, but many antibiotics we have in our arsenal – known as 'superbugs', or multi-drug-resistant (MDR) bacteria. Even where bacteria still have some windows of vulnerability, it can be expensive and time-consuming to identify which drugs will still work – time that people suffering from acute, life-threatening infections simply do not have. In addition to MDR bacteria, there are XDR (extensively drug-resistant) and PDR (pan-drug resistant) bacteria. These bacteria are virtually unstoppable.

Many have been warning about this crisis – officially known as antimicrobial resistance – for decades, alongside an oft-cited projection that globally, up to ten million people could die from these infections every year by 2050. But few realised the crisis was so acute already. The crisis threatens to return us to the pre-antibiotic age, where common illnesses, food poisoning, basic surgical procedures and even an infected cut could develop into life-changing infections, chronic disease, disfigurement, and death.

This has been a silent pandemic that started long before viral threats like COVID and AIDS – not making evening bulletins or closing schools, but now killing millions of people every year. Without new ideas, we face a return to the horrors of a world cowed by bacterial disease, when the crack of a broken tooth or pang of a stomach cramp are once again harbingers of an infection that cannot be stopped.

If we could design the perfect antidote to this crisis, what would it look like? It would be something that kills bacteria efficiently and quickly. It would be cheap and easy to produce from an abundant natural source. It would be easily stored, shipped and administered, with few side-effects or interactions with our own cells. It would target only the bacteria causing us disease, leaving the useful bacteria in our bodies unharmed. It would be easy to adapt or change or evolve as new resistance emergences. And it would complement existing antibiotics, extending the use of the antibiotics we already have.

Bacteriophages are all of these things.

In 2010, Lilli Holst, a South African university student, was taking part in an exciting international project to help students find and isolate their own phages from their local environment.* Drawn to her parents' compost heap, Holst decided that a half-buried, decomposing aubergine was the perfect place to scrape a sample from.

Almost a decade later, a phage isolated from this piece of rotting veg – which Holst had named 'Muddy' – was injected into the arm of seventeen-year-old Isabelle Carnell-Holdaway at London's Great Ormond Street Hospital. Isabelle was fighting an aggressive drug-resistant *mycobacterium* infection that had taken over her body following a double lung transplant; she had been given just a 1% chance of survival. Aggressive treatments with a variety of antibiotics had failed to halt the progress of the infection, which was soon so advanced that festering purple pockets of bacteria were pushing up through the skin on her arms and legs. With nothing to lose, three different types of viruses, including Muddy, were injected into Isabelle's bloodstream twice a day and slathered directly onto the lesions on her skin. Within days, her condition had stabilised. Within

---

\* The project, known as SEA-PHAGES (Science Education Alliance–Phage Hunters Advancing Genomics and Evolutionary Science), has so far helped identify over 20,000 new phages. We'll hear more about it on p. 296.

weeks, the skin lesions were disappearing and wounds that had been open for months were beginning to close up. A few months later, Isabelle was able to leave the hospital.

Muddy has since been used in over a dozen similar cases, and the virus from the rotten aubergine has proven to be one of the most useful phages for this particular type of infection. Sadly, Isabelle died in 2022 from complications of her multiple health challenges. Yet this random virus, found in a random place, had enabled her to live several years, allowing her to learn to drive, travel abroad and tick several other items off her bucket list before she died.

As we will see in this book, phage therapy can be used to save lives, but it is far from simple. The number of people able to benefit from these remarkable viruses remains tiny, and many of those who do receive phage therapy receive it as a last resort, far too late in their illness. It has proven difficult to conclusively prove to investors and regulators that these unusual treatments work in traditional clinical trials. Despite repeated attempts, modern Western medicine just can't seem to get its collective head around the idea of making people better with bacterial viruses.

Those trying to make it work are hampered by scepticism and prejudice because of the wild, wacky and Stalin-tainted history of this one-hundred-year-old idea – as well as regulatory systems built up over many decades to assess the safety of pharmaceutical-grade chemicals, not viruses dredged from compost heaps or found in human waste. With a treatment this complex, which can't be patented and often must be tailored specifically to each patient, the challenge of making phage therapy workable at scale and affordable for all, not just for a lucky few, is perhaps even greater.

As awareness of their power to kill bacteria grows, phages are now finding use in everything from deodorants to probiotics for penguins. They are being used to save crops, pollinating bees and even valuable silkworms from bacterial infections, and as tiny sensors of bacterial contamination. Yet still, the battle to use them in mainstream human medicine continues. And so hundreds of people with chronic or potentially fatal drug-resistant infections still must make the difficult trip to Georgia for treatment using phages in poorly equipped clinics. Many, many more – millions perhaps – don't even get that chance. Even if they or their doctors have heard of phage therapy, they certainly wouldn't know how to access it.

The situation is changing, however, albeit slowly. As the antimicrobial resistance crisis bites, nature's original antibacterials are once again the focus of huge amounts of interest and investment. Where once regulators and funders turned their nose up at the idea of using phages in medicine, they are now open to specially adapted regulatory systems and clinical trials. Research to catalogue and classify the vast unknowns of the 'phageosphere' is finally beginning in earnest, and more scientists are beginning to recognise the importance of phages to our general health. Science journalists like me are breathlessly writing books about them.

Those who study phages know their time in the spotlight is long overdue. In this book I hope to celebrate all that phages have done for us in just a century, and what they could do for us in the next hundred years if given the attention they deserve. We just have to see them for what they are – an essential and necessary part of life on this planet. Phages are the 'dark matter' of life on Earth, central to the evolution of more complex life and the continuation of every ecosystem on the planet but largely unclassified and unstudied. They are our potential allies in the critical global battles against bacterial contamination and infection. And they are front and centre of some of the most exciting and important science and technology of our time.

Still, too few people have even heard of phages, or the weird and wonderful people who have tried to alert us to their importance over the years. I hope this book goes some way to changing that. This is the story of our remarkable world of phages, and a century of struggle to get the world to see them.

# PART 1

*First Phages*

# 1

## Something in the water

Starting in the fresh, still air of the western Himalayas and washing into the hot, tiger-filled mangrove swamps of Bangladesh, the Ganges rushes and rolls and ripples through the towns and cities of northwest India, one of the most densely populated regions of the world. The river throngs with life, death and everything in-between. As well as being home to dolphins, otters, turtles, fish and gharials, the 2,500km river is thought to support a whopping 11% of all the people on the planet. Every twelve years, tens of millions of pilgrims congregate in the towns of Allahabad and Haridwar to bathe together for the *Kumbh Mela*, the world's largest religious festival and possibly the largest gathering of humanity on the planet.

The Ganges is also buzzing with some of Earth's most dangerous bacterial life. Its water contains up to ten times more faecal bacteria than the Indian authorities' 'safe limit' and frighteningly high levels of pathogens that are resistant to anti-biotic drugs.[9] Despite this, the river has long been believed to be capable of healing the sick and even to be somehow self-purifying. In the Hindu faith, the Ma Ganga, or the mother Ganges, is a divine entity formed from the Milky Way, flowing from heaven to cleanse humanity of its sins.

As well as discarded idols, flowers, food waste and the ubiquitous plastic trash that appears around all modern humans, the pewter-coloured river flushes out all manner of waste from the vast region around it: millions of tonnes of sewage and wastewater; dyes and solvents from tanneries and garment factories, fertilisers from nearby farms; animal dung, animal offal and animal carcasses; debris from endless construction and demolition and human remains.

In the holy city of Varanasi, piles of smoking ash from funeral pyres burning twenty-four hours a day wash off the banks of the river. During the peak of the COVID-19 pandemic, in 2020, hundreds of partially cremated bodies could be seen bobbing down the Ganges, and many more sank into its sandy banks. Surely, even a holy river must have a limit to what it can purify?

Yet, somehow, the Ganges' association with cleansing and healing remains. Many swear that bathing in its water helps cure, not cause, infectious diseases. In the nineteenth century, it was observed that when epidemics of infectious diseases like cholera broke out around the Ganges, they often didn't spread downstream, as would be expected at places where wastewater and sewage enters the water regularly. Why? What could possibly be in the water of this famous river that could counteract the effect of trillions of hideous bugs finding their way into it daily?

In 1892, an English scientist called Ernest Hanbury Hankin set sail for India. With the tips of his twirled moustache extended as wide as his ears, the Cambridge microbiologist intended to study cholera and other infectious diseases in the country. It was an era in which such illnesses were still believed by many to be spread by 'miasmas' or putrid air, and Hankin's subsequent suggestion to boil dirty drinking water to help prevent illness was viewed as a rather novel idea.

After conducting a series of tests on water from the Ganges, Hankin found something extraordinary. Water from the river did indeed seem to have an antibiotic effect, killing cultures of the water-borne bacteria, *Vibrio cholerae*, that causes cholera. Hankin noted: 'the unboiled water of the Ganges kills the cholera germ in less than three hours. The same water, when boiled, does not have the same effect.'[10] There was some invisible essence in the murky waters of the Ganges that seemed to be helping limit the spread of cholera in the region.

Hankin's observation, published in the annals of the famous Pasteur Institute in 1896, is thought to be the first description in print of what was probably the bactericidal action of phages.

It has since been found that the Ganges' extraordinary abundance of bacteria encourages a similarly extraordinary abundance of the viruses that prey on them, and that epidemics of cholera are closely linked to bacteriophage numbers.[11] When phage numbers are low, the *Vibrio cholerae* bacteria start to proliferate, causing outbreaks of disease. With lots of hosts available to replicate in, the phages then explode in number, causing the epidemic to collapse as the phages start to wipe the bacteria out. Then, with all their hosts gone, phage numbers fall again, starting the cycle anew.

In 2016, Indian scientists studying the river found there were phages capable of destroying human pathogens even at the Ganges' source, high up in the Himalayas and far from human activity.[12] Here, in a special region known as Gomukh, melting permafrost gathers from a trickle to a stream to a gushing torrent, and the Ganges begins its epic journey seawards. The authors of the study suggest viruses released from this ancient sediment could be the 'seed' that gives the Ganges its especially high levels of phage – and its revered healing properties. That, and the vast amount of our murky wastewater and sewage – packed with phages – that has flowed into the river each day for thousands of years. The Ganges' holiness is, as the scholar Rijul Kochhar puts it, a result of the remarkable confluence of 'faith, filth and phage'.[13]

After Hankin's remarkable findings in India, a variety of scientists witnessed similar phenomena elsewhere. In what has been called the 'prehistory' of phage science, scholars have identified up to thirty scientific observations from the nineteenth and early twentieth century that may be unknowingly describing phages killing bacteria.[14] A few years after Hankin, for example, a Ukrainian bacteriologist called Nikolay Gamaleya published an article in which he described an unknown agent, produced in a dying culture of anthrax bacteria (*Bacillus anthracis*), which seemed able to destroy other cultures of the same bacteria.[15]

Countless microbiologists throughout history may have observed cultures of bacteria that suddenly died, or grew patchily during their work, and many would have put it down to a failure of their technique or simply the vagaries of working with microbes. (And maybe cheesemakers, brewers and any number of people throughout history who worked with microbes too.) Few would have considered that a bacterial virus may be responsible. In Hankin's time, virtually nothing was known about what viruses were or how they caused disease, and they were of course far too small to be seen with even the most powerful microscopes of the era.

The term 'virus' derives from a rather vague Latin term with meanings including 'poison', 'sap of plants', 'slimy liquid', and 'a potent juice', and the Romans used the term for everything from snakes' venom to semen. In Middle English, the word increasingly came to mean a putrid substance that might spread disease. 'A kind of watery stinking Matter, which issues out of Ulcers, being endued with eating and malignant Qualities' is a rather fun definition from a dictionary in 1770.[16]

In the nineteenth century, the pioneers of vaccination, Edward Jenner and Louis Pasteur, realised that for infectious diseases without a bacterial cause there might be a pathogen present that was far too small to detect. They developed vaccines against viral infections but still did not really understand the nature of the infectious agents themselves.

By the 1890s, two scientists, the Russian Dmitri Ivanovsky and Dutch Martinus Beijerinck, had independently deduced that a disease of plants – known as tobacco mosaic disease – was caused by an infectious agent smaller than a bacterium. They took water from plants infected with the disease and put it through what is known as a Chamberlain Filter – a chamber where water is forced through the ultra-fine pores of unglazed porcelain under pressure. This filter could remove even the smallest bacteria from a liquid. Ivanovsky and Beijerinck showed that the filtered and apparently microbe-free water could still cause tobacco mosaic disease when sprayed on plants. Beijerinck called this *Contagium vivum fluidum*, essentially classifying it as a 'contagious life form with the properties of a liquid'. Aside from this work, nobody knew what viruses were. Even as the twentieth century dawned, the term 'virus' was still generally used to describe an infectious disease with a cause that could not be seen or explained.

Then, in the mid-1910s, our understanding of viruses took a dramatic turn. Phages – having gone unseen for the whole of human history – suddenly appeared in visible form to two different scientists, in two different countries, at almost exactly the same time, scorching perfect little holes in plates of bacteria like smoke signals rising from the world of the small. It would start an almost unprecedented period of scientific bickering that would go on for several decades.

# 2

## A microbe of a microbe

Frederick Twort was born in 1877, in the leafy southern English county of Surrey. The son of a strict country doctor,[17] Twort began studying medicine at just sixteen at St Thomas's Hospital in London and, after excelling in his studies, accepted the first paid research post he was offered. 'It was imperative that I earn at least sufficient to be able to pay for lodgings and food,' the sober and immaculate-looking Twort wrote of his early career.[18]

Later in life, Twort would gain a great scientific rival in the form of eccentric French-Canadian microbiologist Felix d'Herelle, who had a rather different approach to his scientific education. Four years older than Twort, d'Herelle had also left school at sixteen, but spent his late teens travelling around Europe and South America, attending talks and lectures when it took his fancy, reading books and spending his well-connected family's money.[19]

The two men grew up in an era when 'microbe hunting' was a glamorous profession that had captured the world's attention. Bacteriologists like Louis Pasteur, Robert Koch and Paul Ehrlich had become household names as their work advanced understanding of the causes and prevention of contagious diseases, and the public and politicians alike eagerly awaited the latest discoveries, conjured from flasks of putrid broth in the light of a gas flame. In France, Pasteur was virtually a saint. Not only had he developed a range of vaccines and serums to prevent illness, but he had also developed the miraculous process of 'pasteurisation', which finally allowed people to prevent bacterial contamination of everyday items like food, milk, water and wine. It is still widely used in food production today. But in a world without any true antibiotics – that is, drugs that reliably killed bacteria in the body without side effects – infectious disease still played a huge part in the lives of everyone, rich or poor, young or old. The average age one could expect to live to in Europe at the turn of the twentieth century was just over forty – even lower in poorer countries.

D'Herelle idolised Pasteur, and would later claim in his memoirs that his entire life plan from an early age was to emulate the career path of the great French scientist.[20] Considering that Pasteur attended one of the most prestigious medical schools in France, and d'Herelle left school with few formal qualifications, he set about it rather oddly.

Nevertheless, d'Herelle saw himself as a potential star microbe hunter of the future. On one of his plentiful teenage holidays, cycling around rural France and Belgium, he overheard locals talking about a boy who had been bitten by a rabid dog and the plan to rush the boy to a nearby abbey, where monks could apparently cure the terrifying disease. With little else on his schedule that day, the precocious sixteen-year-old cycled 60km to the abbey to find out more about the monks and their proposed treatment, which mainly involved a lot of chanting and a bit of old thread from the clothing of an eighth-century bishop. D'Herelle was intrigued and studiously compared the monks' alleged success rate with that of a rabies vaccine being tested by Pasteur in Paris. He had performed his first scientific review, of sorts.

'It is probable that I have, by birth, the first required quality needed to make a good microbe hunter,' he wrote after another encounter with disease and death.[21] While travelling back from a holiday in South America, an outbreak of yellow fever struck the ship he was on shortly after leaving Rio de Janeiro. 'One morning, very early, seven bodies, one-by-one, slid into the sea . . . Most of the passengers were in anguish: I was perfectly calm, I thought I was invincible.'

He continued to travel for years, staying in England, Greece and Turkey, where he met his future wife, Marie Caire, who was just fifteen.[22] According to d'Herelle's biographer William Summers, the couple married quickly and enjoyed a 'leisurely life of idleness'. At twenty-four, d'Herelle finally decided to settle into a career, and moved to Canada, where there were so few microbiologists that he simply declared himself one and set up his own home laboratory. In his first scientific paper, he claimed that he had discovered carbon was 'not an element' – dismissing the work of the great chemists of the eighteenth and nineteenth centuries – following an experiment he conducted on radishes.[23]

Back in England, meanwhile, Twort's scientific career had progressed along a more conventional path. After his medical degree, he studied under the noted bacteriologist Professor William Bulloch, published his first paper aged twenty-eight, and a few years later gained a permanent research post at the Brown Institute, a veterinary hospital and research faculty in Vauxhall, just south of the River Thames.[24]

At about the same age, his wandering French-Canadian peer had only just settled on what to call himself. Born Hubert

Augustin Félix Haerens, he was known at various points as Félix Hoerens and Hubert Félix Haerens d'Hérelle, before he finally settled on Felix d'Herelle in around 1901. There is also some debate over where he was actually born. D'Herelle's claim to have been born in Canada seems to be contradicted by a Parisian birth certificate unearthed in the early 2000s.[25] Some historians believe he may have been Belgian.[26] His shape-shifting identity may be explained by the fact that by his early twenties, he had presided over a bankrupted chocolate factory and was listed as 'deserted' from the French Army.[27]

Photos of a young d'Herelle show a roguishly handsome and even slightly villainous-looking man, like some hard-drinking saloon owner from the Wild West. His dark hair was buzzed short, and a Lenin-style goatee was sometimes topped with a flamboyantly twirled moustache. But the real drama was in his eyes – you could carry your shopping in the dark bags under them. In the rare images where he looks directly into the camera lens, he stares out from tired, guilty peepers. Later in life, the dark hair receded and turned grey, the goatee became bushy and the chain-smoking microbiologist swapped the dapper suits and skinny ties for a bulging full-length lab coat – but the dark and doleful eyes remained.

Twort – with his carefully parted hair, often photographed sporting a gentle smile in his British Army officer's uniform – was the yin to the wild-eyed yang of d'Herelle. In his Brown Institute lab, crammed beside the busy London and South Western Railway and a smelly Beef Tea factory, he invented a new type of staining to reveal the fine structure of microbes, and made important progress in his research focused on bacteria that caused wasting disease in cattle. He liked to work alone.[28]

Meanwhile, across the ocean, the French, Canadian or possibly Belgian adventurer eventually secured his first semi-official scientific job with the help of his well-connected father. In a strikingly Canadian turn of events, the Quebec government employed him to see if it was possible to turn its excess maple syrup into whiskey. In his DIY laboratory, he set about using his self-taught microbiology skills to turn the sticky sap into a liquor that could be sold to the United States. It didn't work. Unfazed, he saw an advert for a government scientist in Guatemala, a newly independent Latin American state, and yearning for more travel, he applied. To his surprise, he got the job, and his career as a microbiologist began. It was a country where few questions would be asked about his lack of qualifications.[29]

As Twort was diligently building a research career in London, d'Herelle was living on the edge in the Americas. Guatemala, in the late 1800s, was a wild, lawless place, and he was soon doing wild, lawless science. Without any formal training, d'Herelle found himself responsible for bacteriological examinations of patients at the General Hospital in Guatemala City, and even acted as an examiner for students. Tasked with dealing with an outbreak of yellow fever, he apparently burned a row of family houses to the ground when his quarantine orders were not obeyed. He had been advised on arrival to carry a revolver when outside of the cities, and so when a drunken man staggered up to him wielding a knife, he discharged his revolver into his chest, piercing his heart and killing him.

'After dressing my wound in a so-so fashion (I still carry the scar), I continued my trip,' he wrote of the incident. 'At that time, in such cases, it was preferable not to have dealings with the authorities; I certainly had nothing to worry about, but I did not want to be delayed for some time, under the pretext of an inquest.'[30]

His wife Marie and their two young daughters, the youngest only eighteen months old, were not spared from the madness. Plagued by heat and humidity and by new and unfamiliar toxic plants and animals, all three contracted malaria repeatedly, as well as yellow fever. By 1907, d'Herelle had been hired by the government in nearby Mexico (again, as he wrote later, 'where anyone could be a doctor, no degree was required') and his family were so badly affected by diarrhoea and an intermittent water supply that they at one point looked like skin and bone. 'We are all very depressed', wrote his long-suffering wife Marie.[31]

Twort and d'Herelle could hardly have been more different in terms of career path and character. Yet while they worked on different sides of the world, on different questions, with different bacterial cultures, both noticed a certain strange phenomenon which would later define both their careers and bring their paths crashing together.

Twort, among the dark wood, brass and chequered tiles of his hospital lab in England, had been trying unsuccessfully to work out how to grow human and animal viruses in the lab so that they could be observed and studied like bacteria. With so little known about viruses at the time, Twort would certainly not have known that viruses require the facilities of a living cell to replicate and so his attempts to grow the smallpox virus on various chemical extracts, bacterial mixtures and nutrient broths were doomed to fail.

He tested hundreds of different mixtures, none of which worked as a growth medium for the virus he wanted to study, smallpox. But he did notice something odd in one of his plates, where instead of virus, bacteria had started to grow. While less attentive scientists might have cursed another batch of contaminated plates and sent them for cleaning, Twort observed them carefully: one bacterial colony appeared itself to be afflicted with some kind of disease. On the otherwise uniform film of bacterial cells, there were tiny 'glassy' holes, or plaques, where the bacteria had simply vanished. The holes grew, quickly, leaving behind nothing that could be seen under the microscope except 'fine granules', as he described them later.[32]

Intrigued, Twort switched the focus of his research to investigate what might be causing these holes or plaques. He took a sample from the middle of a plaque, diluted it, and passed it through the same ultra-fine porcelain filters used by Ivanovsky and Beijerinck to remove all traces of microbes from the liquid. After adding the filtered substance to another plate of bacteria, he saw the same 'glassy transformation' of healthy bacterial cells into regions of empty space. The point where he touched the substance to the bacteria soon became transparent, and gradually made the whole colony transparent. When he examined the transformed areas, he found 'minute granules' again but no sign of any bacteria, where just hours before they were packed solid.

He could in fact repeat this process seemingly indefinitely, the same glassy transformation spreading out each time he transferred a touch of it to a fresh plate, even after diluting it by a million-fold. Whatever was causing the spots passed through a porcelain microbe filter as if it were a liquid. But it also grew forth, endlessly, from the minutest contamination, like a microbe.

The outbreak of war in Europe, in 1914, effectively ended Twort's ability to study this odd phenomenon. Like many others, his research funding was cut and he joined the Royal Army Medical Corps in late 1915 to run a wartime medical lab in Greece.[33] He published his observations of the strange spots in the famous medical journal *The Lancet* that year, buried halfway through a report on his failed attempts to grow small-pox virus. In his meandering article he wondered if the 'dissolving substance' was a 'separate form of life'. But crucially, he stopped short of making any concrete claims about what was causing the holes on the bacterial plates.[34] He concluded his paper cautiously:

> From these results it is difficult to draw conclusions . . . this may be living protoplasm that forms no individuals, or an enzyme with the power of growth . . . In any case, whatever explanation is accepted, the possibility of its being an ultra-microscopic virus has not been definitively disproved because we do not know for certain the nature of such a virus . . . I regret that financial considerations have prevented my carrying these researches to a definite conclusion, but I have indicated the lines along which others more fortunately situated can proceed.

In other words: this strange stuff I grew could be a virus. But we can't say for sure, because we don't really know what viruses are. And now I've run out of money.

A few years before Twort published his paper, d'Herelle made his first major breakthrough as a scientist in Mexico. When large plagues of locusts hit the Yucatán region, the country's southeastern peninsula, the ever curious d'Herelle developed the idea of using the locusts' own diseases to help farmers wipe them out. Still the oddball who had followed a rabid child to an abbey on his bike, he travelled the country in search of dead bugs and encouraged people to send him expired insects. He eventually isolated a bacterium that inflicted locusts with a nasty black diarrhoea, grew it in large amounts and repurposed it to sprinkle on crops, thus creating a disease epidemic in any swarm arriving to feed. Like any good microbiologist of the era, he had tested the safety of his biological pesticide by ingesting it himself, after testing it first on dogs and sheep. He was soon invited to test his idea on crops in Argentina, where the insects were causing untold damage. It worked, with the pests falling from the sky days later, and dead locusts found up to 20 miles from the infected fields.

In 1912 *The New York Times* announced: 'French Doctor's campaign of extermination a complete success.'[35] Although, of course, he was not actually a doctor, he may not have been French and some would later argue his methods were not as successful as he claimed. The idea of using nature itself to fight nature – an epidemic to fight a plague – was really rather pioneering, and would become a feature of d'Herelle's work. His career gaining momentum, it was around this time, during one of his frequent trips back to France, that he managed to wrangle himself a position as an unpaid assistant at the world famous Pasteur Institute in Paris – the spiritual home of his childhood hero. When World War I broke out in 1914, he helped produce vaccines for the war effort and was promoted to a laboratory chief.[36]

Whereas the outbreak of war had stopped Twort from progressing any further with his work, for d'Herelle it provided the backdrop for his greatest moment. In 1915 he was sent by the Pasteur Institute to investigate an epidemic of severe dysentery in French troops stationed at Maisons-Laffitte, on the outskirts of Paris. Dysentery – an umbrella term for intestinal infection and inflammation – was one of a number of so-called 'war diseases' causing untold damage to armies all around the globe. It was often caused by *Shigella* bacteria and could spread voraciously among anyone living in unsanitary conditions. The ultimate acute gastrointestinal illness, it causes crippling stomach cramps, diarrhoea, then bloody diarrhoea, leading to extreme dehydration, emaciation and, eventually, a painful and undignified death.

D'Herelle, no stranger to a watery stool, dived into his work, collecting the excretions of the ill soldiers and declaring the unusually nasty strain of bacteria in this particular outbreak should be named after him. Over the next year, on top of his work making vaccines, he continued to investigate the microbes he had obtained from the poo-cocktails taken at the barracks in any spare time he had. And this is when he hit upon the strange idea that would define his career: that it might be possible to destroy the nasty *Shigella* microbe with an 'ultra-microbe' that itself attacked microbes.

While working with the locusts in Mexico – which were suffering from the insect equivalent of dysentery – d'Herelle had noticed something odd. While culturing the locust-killing bacteria on plates, he'd seen the same mysterious, glassy spots as Twort, as if some kind of disinfectant substance was growing on his bacterial samples. Again like Twort, he found that a tiny sample from the spotted area could be wiped on another plate of bacteria, and more spots would appear on that plate too, sometimes just a few hours later, even when the substance was diluted many times.

Later, d'Herelle would claim he saw these spots as early as 1910, long before Twort's publication in 1915. But crucially, he never published his observations at the time. Regardless, what both men had seen was phages in action – a viral outbreak causing an epidemic of death in a bacterial colony. After remaining unseen for so long, phages had suddenly been noticed by two different scientists virtually at the same time.

For Twort, it was not immediately obvious what these small circles were telling him, and circumstance prevented him from continuing his studies. But by 1916, d'Herelle had struck upon the idea of trying to harness the power of this mysterious bacteria-killing phenomena in the fight against dysentery. And unlike Twort, he was not backwards about coming forward with a theory about what was happening on his plates – he believed he had found a new kind of microbe that could help us fight off bacterial disease.

Now in his forties, perhaps he felt a pressure to come up with a big idea – perhaps to keep his 'life plan' on track, or perhaps just to keep the questions about who he was, and his right to be working as a scientist at the Pasteur Institute, at bay. He roped in his wife and young daughters to help with his day job at the Pasteur Institute, allowing him to spend longer working on this mysterious substance that killed bacteria in the evenings. To try and prove the hypothesis he had been work-ing on, d'Herelle devised what is described by his biographer, William Summers, as 'an incomprehensible procedure'.[37] He passed the soldiers' poo through the ultra-fine porcelain filter to remove the microbes, then added this highly filtered liquid back into a flask full of the dysentery bacteria. He also added a small amount of the filtered poo onto colonies of dysentery bacteria growing flat on glass plates. He put them into an incubator and retired, at last, to bed.

Thanks to the colourful graphics of viruses you see on television and in newspapers, you probably think of a virus as some mean-looking bug that drifts around waiting to infect something – you might be picturing a spiky red ball, like a coronavirus, or the wonderful T4 phage, with its angular head and long spiky legs. But these things aren't really viruses. These are just the virus's transportation device, a 'virion', carrying around the all-important yet dormant chemical code, and helping the code survive in the world long enough to reach a new host. This thing may have a distinct form and shape, sometimes with a head and a tail and arm-like structures, but it is not alive in any real sense. Yes, it is poised to recognise and bind to a potential host cell, but otherwise it is completely inert and lifeless, drifting through the world like a seed, or a spore.

The virus only truly 'comes to life' when the virion latches onto a suitable host cell and the sleeping genetic code is injected into its prey. The husk of the virion stays abandoned outside, like a spent rocket. Only then can the virus begin to exert some kind of effect on the world. If the virus can evade the host cell's defences, the complex molecular machinery that the cell uses to read and act upon the information in its own genes will begin to read the virus's genes too. The cell has been hijacked and is soon manufacturing the proteins and other biological molecules required to make new virions at the expense of its own health. The bacterium has become a virion factory: its metabolism focused on making phage head proteins, phage tail proteins, the receptor molecules that help the phage attach to potential hosts and hundreds of copies of the phage's genetic code.

And now for the really neat part: the components spontaneously self-assemble, pushed and pulled into shape by the tiny attractive and repellent forces of the molecules they are made up of, while other special proteins encoded in the phage's genes act as scaffolds or guide the various components into place. A molecular motor at the top of the tail packs the long length of DNA – which doesn't naturally want to coil or fold into a tight space – so tightly into the phage's heads that scientists estimate the pressure inside them to be thirty times higher than a car tyre.[38] It's both a meticulous and miraculous process, like a giant bin full of doll parts all suddenly sticking themselves together into complete models.

Once the cell is fit to bursting with new virions, the original virus issues its final devastating genetic instruction to the hijacked cell: to metabolise 'suicide molecules' – the enzymes that will burst the cell open from the inside, releasing all the new virions into the world to find new hosts and repeat the process. Because bacteria maintain a relatively high internal pressure relative to their surroundings, the cell's final moments can be explosive. The chemical countdown that leads to this explosion is so exquisitely calibrated that scientists can predict, almost to the second, when an infected cell will go 'poof', the tough circular or rod-shaped cells visible one moment, then a mess of burst membranes and splattered insides the next.* The number of progeny produced by one infected cell can range from a few dozen to tens of thousands, depending on the virus. But only one needs to be successful for the lineage to continue.

Remember: this is going on all around us, on us and inside us, every second of our lives. Our bodies are home to trillions of bacterial cells and therefore even more bacteriophage virions. So are the environments in which we live. The entire biosphere is positively humming with viruses bursting forth from bacterial cells. Thankfully, all these microbial predators rarely cause us any harm. And in fact, as d'Herelle would soon discover, the deadly efficiency of phages can even be rather useful.

While d'Herelle was sleeping, a battle as old as life itself began to play out in the flask and plates in his incubator. The cocktail of poo he had collected from the sick soldiers originally contained many different types of bacteria, including those responsible for the awful symptoms of dysentery. But they had all been removed by the fine bacterial filter. Virions, being far, far smaller than bacteria, however, passed through the filter and remained. The mixture was now clear to the naked eye but still contained a rather horrid cocktail of any virions that happened to be lurking in the soldier's guts that day. If the mixture did contain phages that specifically attacked *Shigella* bacteria, then

---

* The extraordinary proteins that enable phages to do this (produced unwittingly by their hosts) are now being studied as potential antibiotics of the future.

adding this mixture back into pure cultures of *Shigella* bacteria would essentially light the touch paper to a microbial war. Not all the virions in d'Herelle's mixture were ones that target *Shigella* bacteria, however; there were likely many others in there, animal, plant and human viruses, plus phages that target other bacteria.* These, when added to the *Shigella* cultures, would simply bounce around harmlessly, hopelessly waiting to encounter whatever host cell they need. But a virion of a phage that targeted *Shigella* – that would have a field day, suddenly surrounded by an abundance of suitable cells to prey on.

By the time d'Herelle popped on his nightcap, some *Shigella* bacterial cells may well have been infected with a *Shigella*-loving phage. The phage virions would bind to the surface of their hosts, puncture the membrane and inject their genomes inside. Within as little as half an hour, the cell would have burst, and hundreds of identical phages would have spilled out into the mixture. The surrounding *Shigella* cells would be quickly infected with *Shigella*-loving phages too and would themselves soon burst forth with hundreds more phages, unleashing wave after wave of more predators.

The flask, once opaque with floating *Shigella* cells, would clear to a transparent gloop of broken bacterial bodies and a trillions-strong army of invisible *Shigella*-phage clones – so small that they do not reflect or absorb light. The plates, where there was once a dense healthy lawn of *Shigella* bacteria, would resemble a microbial killing field, covered in holes where the tiny epidemics were spreading.

---

* Recent work by Dr Ellie Jameson at Warwick University found that the most abundant type of virus in the human gut can be viruses of peppers, depending on the diet of the population.

Thanks to the elegance of d'Herelle's experimentation, the age-old battle between phage and bacterium was played out in its purest, most observable form; not within the black box of the human gut or at the bottom of a murky river, just one visible colony of bacteria being set upon by its viral nemesis in a clear glass flask. Just one or two phages dropped into these massed populations of bacterial cells, and left overnight, the next day could have become a thousand trillion. D'Herelle's recollection of his discovery the next morning seems to suggest he understood perfectly what had happened overnight, but also that what he had seen might change his life:

> On opening the incubator, I experienced one of those rare moments of intense emotion . . . I saw that the broth culture, which the night before had been very turbid, was perfectly clear: all the bacteria had vanished, they had dissolved away like sugar in water. As for the agar spread, it was devoid of all growth . . . in a flash, I had understood: what caused my clear spots was in fact an invisible microbe, a filterable virus, but a virus parasitic on bacteria.[39]

Apparently unaware of Twort's cautious observations published a year earlier, d'Herelle believed he had discovered something entirely new to science. Like Twort, he found one tiny sample from a spot, diluted many times, could then be added to more pure bacteria and the same effect would occur, over and over. Were it simply some kind of antiseptic liquid, there would be a diminishing effect with each dilution. Instead, whatever was present was reproducing, seemingly at the expense of the bacteria, just like the 'filterable viruses' that infected tobacco plants or people. The possibility that a bacteria-killing virus could find use in human health was immediately obvious to d'Herelle. He began to wonder if phages play a role every day in helping to prevent and resolve bacterial illnesses:

> Another thought came to me also. If this is true, the same thing will have probably occurred in the sick man. In his intestine, as in my test-tube, the dysentery bacilli will have dissolved away under the action of their parasite. He should now be cured.

After his flash of inspiration, d'Herelle broke the news to his long-suffering family, who helped him decide on the all-important next step: giving his newly discovered life form a name.

> In the evening, under the light, I was telling my loved ones what I had seen: the dysentery bacilli devoured by a 'microbe of microbes'. My wife asked me, 'what are you going to call them?' And the four of us put our heads together. Name after name was suggested and discarded again. Finally . . . we came up with the word 'bacteriophage', a word formed from 'bacter-ium' and 'phagein', the Greek word for 'eat'.

D'Herelle's intriguing name stuck, and all viruses of bacteria are still to this day known as 'phages'. D'Herelle's desire for a catchy name sets bacterial viruses apart from all other viruses – those that affect plants, or animals, or fungi, or humans – which have all remained simply 'viruses'. In a note presented to the French Academy of Sciences in 1917 entitled 'On an invisible microbe antagonistic to dysentery bacilli,[40] d'Herelle showed none of the polite caution of Twort. In just two pages, he boldly explained that he had discovered a new form of life.

By 1918 he published an elegantly simple experiment to prove his substance was not a liquid with the power of growth, but tiny particles or microbes. If it was liquid, when a solution of the heavily diluted substance was washed across a bacterial plate, its effect would surely be uniform across the whole area of the bacterial cells. It wasn't – spots appeared here and there, randomly. D'Herelle said this must be evidence of individual viruses landing on individual cells, multiplying and infecting more and more cells around it until a circular spot – a crater of destruction and death among the bacterial lawn – could finally be seen with the naked eye. This technique even allowed d'Herelle to start to quantify the number of virus particles he was dealing with; by counting the spots and working backwards through the dilutions of the solution, he could work out how many individual phages were in his original sample. The elegantly simple techniques he developed for detecting and counting phages remain, to this day, the standard way to find and isolate bacterial viruses.

D'Herelle continued to present his findings through notes submitted to the French Academy of Science. But remarkably, the response was generally indifferent. 'Opinions on d'Herelle ranged from visionary to fool', according to the science historian and phage expert Alain Dublanchet. And another one of d'Herelle's peers stated: 'If this bacteriophage really exists, during all the time I have been a bacteriologist, I would surely have observed it.'[41]

The scientific establishment were not just miffed that they had missed such an important new microbe. For as long as d'Herelle had been known as a scientist, a dark cloud had hung over his name. There were rumours he had exaggerated his successes in South America, and forged results in France. There were doubts about his character and trustworthiness. He was described as 'annoying' by one colleague and 'hypersensitive' by another.[42,43] By the late 1910s, it was likely that his colleagues at the Pasteur Institute had discovered he didn't have a medical degree in a role that required one, and one scientist who saw him demonstrate his phage experiments accused him of 'conjuring tricks'.[44] Surely this uneducated upstart couldn't have made the greatest microbiological discovery of the century so far?

Perhaps fairly, d'Herelle felt he was being ostracised simply because his theories made scientists far more senior than him look bad. But it didn't help that everything he wrote bristled with pugnacity against those who came before him or dared to question him, including his own bosses at the Pasteur Institute. D'Herelle's wild and abrasive style only encouraged his peers to prove he was wrong. Various theories were concocted to explain the self-propagating bacteria-killing substance that d'Herelle had revealed to the world, and despite the elegant plaque experiments, remarkably few scientists believed it was truly a 'microbe of a microbe'. The leading alternative explanation was that bacteriophage was some kind of enzyme (or a 'ferment' in the parlance of the time) with 'the power of growth' – perhaps secreted by the bacterium itself. This may have been because other scientists had recently discovered enzymes capable of breaking down bacterial cell walls, including the Scottish physician and microbiologist Sir Alexander Fleming, who found such enzymes in both tears and egg white in 1922.[45]

Others supposed it was genetic material (not far off) and a third group argued it was in fact a self-destructive stage of bacterial life (very far off). At one point d'Herelle roped none other than Albert Einstein into the debate, asking him to review the maths of the experiments showing that phages were individual particles not a liquid. Einstein sided with d'Herelle.[46] But even the backing of the world's most famous physicist did not convince the senior directors at the Pasteur Institute, who continued a campaign of research to try and disprove d'Herelle's results.

D'Herelle increasingly began to see himself as a virtuous underdog and oppressed outsider, fighting against the group-think of a conservative establishment. 'Theories have no impact on me whatsoever,' he wrote in his memoirs, which remain unpublished in the archives of the Pasteur Institute to this day. 'If my results coincide with theories, that's fantastic. Then I'll accept them. If not then I'll discard them, no matter what authority defends them ... That's what has caused me the enmity of the "official scholars" in all countries.'[47]

He knew it would take a much more dramatic illustration of what phages could do to get the ignoramuses in Paris to take notice of his discovery.

# 3

## The great phage feud

In August 1919, a desperately ill eleven-year-old boy arrived at the Hôpital des Enfants-Malades in Paris. He was suffering from such severe diarrhoea that he was dehydrated, exhausted and passing nothing but bloody mucous in his stools. The doctors recognised the sadly familiar signs of bacterial dysentery. But this unsuspecting boy, his name recorded only as Robert K, was about to become a human guinea pig.

Two days earlier, d'Herelle had breezed into the hospital and told the chief of paediatrics, Victor Henri Hutinel, that he had found a way to treat the dreaded dysentery. Hutinel, devoid of any better options, agreed to try the experimental remedy on a patient as long as d'Herelle could prove it was safe. This was long before the establishment of medical regulators like the European Medicines Agency or the Federal Drug Administration, and their requirement for years-long safety studies and clinical trials. D'Herelle's idea of a clinical trial was to drink a dose 100 times as strong as the one he would give to a patient in front of the hospital's doctors.[48] 'I had already been taking large amounts of these types of solutions, and all my family members tried it as well', he wrote.[49] Not wanting to miss out, or perhaps motivated by misplaced machismo, twenty doctors at the hospital took a swig, too. None reported any ill-effects, and so the stage was set for the treatment to be deployed. They just needed a patient.

Robert was admitted later that evening. The morning after, with the doctors still reporting no side-effects from the day before, d'Herelle was permitted to give Robert two millilitres of the bacteriophages he had continued to propagate from the ill soldiers at the barracks. He retired home and hoped the phages would do in the boy's guts what they had done overnight in his flask of *Shigella*. By the next day, the boy's stools were loose, but not bloody and he was soon feeling better. Before long, doctors could no longer detect any *Shigella* bacteria in his stools. Days later he left hospital and returned to the Parisian slums, unaware he was the first human patient to ever receive the controversial treatment that would come to be known as 'phage therapy'.

Knowing that dysentery was known to resolve spontaneously, d'Herelle saved his excitement until he had the opportunity to test the phages on other patients. Eventually, another outbreak of dysentery in a Parisian suburb brought to the hospital three brothers, whose sister had already died, along with a fourth unrelated sufferer. The same dose of phages led to the same result: all four were well enough to go home within a week.

It was still early days – just five patients – yet it's hard to overstate how important this result was. No clinic or hospital in the world at the time had access to reliable treatments for treating acute bacterial disease. This was still eight years before Alexander Fleming's famed discovery of penicillin, the first true chemical antibiotic, and over two decades before the substance would be widely available for use in medicine. There were vaccines available for a few infectious diseases, like small-pox, but otherwise when patients developed a bacterial infec-tion there was little to do but try and kill off the cause with toxic substances such as mercury, bromine and derivatives of arsenic, hoping that the bacteria would die before the patient did.

Perhaps feeling burned by criticism of his earlier work, d'Herelle did not get ahead of himself. He recognised the numbers of the patients involved in this impromptu trial did not offer conclusive proof that the method could be used as a cure. The results of his studies were not immediately published – but nevertheless word had spread about his discoveries, draw-ing other scientists to investigate the possibilities of phage medicine. Consequently, the first published study of phage therapy was by some of his peers, when Richard Bruynoghe and Joseph Maisin[50] found bacteriophages could be used to reduce staphylococcal skin disease in lesions deliberately opened in the skin.

After his successes at the children's hospital in 1919, d'Herelle had retreated to the French countryside. A colleague had mentioned that an unusually potent outbreak of avian *Salmonella* was devastating chicken flocks in the French countryside, and he saw an opportunity to conduct a detailed study of how bacteriophages impact bacterial disease, albeit with animals. Among the poop and feathers of the coops, he was able to monitor the spread of the *Salmonella* bacteria meticulously in the chickens, and the emergence and spread of phages that could kill it off. When phages in cured hens were pooped out, that poop was pecked at by other hens, effectively vaccinating them with phages against the *Salmonella* bacteria.

Combining his two great masterstrokes – seeding disease epidemics in locusts and discovering bacterial viruses – he carefully observed how a second epidemic spread among the clucking subjects: a viral epidemic, spreading through the bacteria that were themselves spreading through chickens. And yet, still, most of his colleagues and many other leading figures in microbiology continued to argue that the bacteriophage phenomenon was some kind of enzyme secreted by the bacteria themselves.

Not content with the idea of extinguishing outbreaks of disease with his viruses, d'Herelle was beginning to develop an even bigger idea. His observations on the wards of the soldiers' barracks and among the chickens suggested that the appearance of dysentery-killing phages coincided with people or animals starting to recover spontaneously from the disease. He began to work on a theory that would cause an even bigger upset among the great and good of the scientific–medical world: that phages might be a natural part of our immune systems, helpful guests that could explain how people sometimes spontaneously recovered from infections.[51] He began to see phage therapy as a way to augment the body's existing population of phages in order to help it overcome infections.

It was an idea that threatened to overturn everything that was then known about how the body's immune system worked, a theory later described by science historians as 'heretical' because of the way it clashed with the emerging theory that antibodies helped the body fight infection.[52] It is only over a hundred years later, in recent years, that scientists have shown that phages play a role in protecting us from harmful bacteria, especially within our intestines, which actively recruit certain phages and even transport them to different parts of the gut. The Australian phage researcher Jeremy Barr, one of the leading experts in this emerging field, now calls phages the body's 'third immune system', providing additional support to our innate and adaptive immune systems and helping to keep a healthy selection of bacterial species in our digestive tract.

'Papa leaves for Saigon', d'Herelle's eldest daughter Marcelle wrote in her diary in March 1920.[53] In her father's version of events, he was unable to resist the lure of an exciting new role at the Pasteur Institute's lab in Indochina, now Vietnam, describing the country as 'the land of my dreams' owing to the countless deadly plagues and epidemics raging through it. But correspondence between scientists at the institute suggest he may have been sent there as a form of exile after his clashes with colleagues in Paris.[54]

Far from the hostility of his French colleagues, d'Herelle received help and funding to develop phage therapy in the Far East and many other regions struggling to contain disease outbreaks, such as Egypt and India.

As he travelled the world's tropical disease hotspots, and his confidence in his theories grew, so too did his outspokenness. In his letters and in his scientific publications he made arrogant proclamations about how science should be done, claiming that only those who studied epidemics in their natural, gruesome settings – like him – would find the answer to nature's riddles. 'You will only master nature when you learn to obey it,' he spat at colleagues far more senior and celebrated than him. He wrote off those who modelled disease outbreaks in labs or gave animals human diseases as 'lazy and fearsome'. His belief in the power of nature to fight nature was becoming dogmatic, and he believed his scientific upbringing in the wilds of Guatemala and Mexico was a 'school of hard knocks' that had made him more daring and fearless than others. He thought the more outrageously putrid the place, the more 'nature' he would find, crawling up the walls and seeping into bedpans; when studying cholera in India, he had insisted on working amid the frightful wards of the paupers' hospital rather than the more sterile hospitals set up for the British and their Indian officers.[55]

While working in Saigon, d'Herelle continued to develop his outlandish theory that phages were a vital element of human immune response. As a result, he began to see all existing vaccines and therapeutic serums as based on unsound theories of immunity, and therefore not just ineffective but dangerous. Never one for diplomacy, he began to denounce mainstream science and medicine, its commercial interests and its 'crimes'. He made particularly disparaging comments about the BCG tuberculosis vaccine, the 'C' of which was named after Albert Calmette, deputy director of the Pasteur Institute and effectively his boss.

The dispute between d'Herelle and his colleagues was now about more than whether bacteriophage was a virus or not. The Pasteur Institute was not merely a research institute; it manufactured and sold millions of vials of vaccines and serums from Paris and its various outposts in French colonies around the world. This maverick was now actively undermining one of the institution's main sources of income. More than that, if d'Herelle's theories about bacteriophages' role in infection control and the immune system were true, this angry autodidact would surely leapfrog the many towering figures at the institute working in disease control and immunity who saw themselves in line for the next Nobel Prize. D'Herelle and his invisible microbe had become a threat to the credibility and the business of the institute, and a threat to the aspirations of some of the most eminent scientists in the world. In a letter to senior Pasteurian scientist Émile Roux, d'Herelle declared himself 'absolutely aware' of the radicalism of his ideas and the concerns of the Pasteur Institute. His observations, however, were clear, he wrote – and 'no authority could hamper the truth'.[56]

D'Herelle returned to Paris from Indochina in 1920 to find his entire laboratory had been assigned to someone else. His phage work was defunded, and some of his former staff had been reassigned to research projects seeking to disprove his theory that the bacteriophage was a virus.[57] The man who had made possibly the biggest discovery in bacteriology of the new century was reduced to working from a single wobbly stool in the lab of one of the few colleagues he remained on good terms with, the French scientist and food writer Édouard de Pomiane.[58]

His two greatest enemies were Calmette, still smarting from d'Herelle's scandalous criticism of his vaccines, and Jules Bordet, director of the Pasteur Institute in Brussels and a Nobel Prize-winner for his work on the immune system.[59] D'Herelle would later write of Calmette: 'Of him I will say nothing: he was my declared enemy, he pursued me with his enmity for the rest of his life.'

Jules Bordet had read about d'Herelle's work and decided to try and find phages that would kill the laboratory workhorse bacteria *E.coli*. Replicating d'Herelle's experiments, Bordet did initially produce a filtered substance that could kill batches of *E.coli*. But then in further experiments, he found the substance could not kill the exact strain of bacteria it was extracted from. Bordet concluded that what d'Herelle called 'bacteriophage' was really a toxic protein released by bacteria to kill other bacteria, but not itself.

The idea that phages were part of humans' immune response, which d'Herelle had published in a 1921 book entitled *Bacteriophage and Its Role in Immunity*, was also a direct challenge to Bordet's life's work, which asserted that antibodies were the key element of the body's immune system. A group of researchers linked to Bordet, who became known as the 'Belgian group', dedicated an entire decade to fighting d'Herelle and his ideas. D'Herelle once even had to sue his own institution to force them to publish his response to one of the group's studies.

Despite the hostility from his own institute, d'Herelle had a knack for attracting great excitement about his work wherever he went. Thanks to commercial interest in his ideas he was able to found a private lab in Paris, the *Laboratoire du Bacteriophage*, which manufactured at least five phage-based medicines for different types of infection, named Bacté-coli-phage, Bacté-rhino-phage, Bacté-intesti-phage, Bacté-pyo-phage and Bacté-staphy-phage – the profits of which allowed him to fund further research. (The laboratory was later acquired by French company L'Oréal.)

He travelled constantly, and his local successes continued to wow local doctors, medical officers and journalists. With little if any regulation of medicines, and few other options available for taming bacterial disease, d'Herelle's fame grew, as did the buzz around the possibilities of phage therapy. The Oswaldo Cruz Institute in Rio de Janeiro began producing anti-dysentery phages in 1924 for use in hospitals all over Brazil and across several Latin American countries, prompting other Brazilian medicine manufacturers to create their own phage products. It is estimated that by 1925, more than 150 papers had been published on the therapeutic use of phages.[60] By 1926, d'Herelle had accurately deduced the life history of a bacterial virus, including the attachment of the phage particle to the susceptible bacterium, multiplication of phages within the bacterial cell and the bursting of the bacterium to set free the progeny virus particles – decades before the technology required to actually *see* any of this was available. And yet his embittered rivals continued to push the alternative view, that all this was caused by some kind of bactericidal enzyme.

There was a good reason so many scientists supported alternative explanations of what phages were – more than just animosity towards d'Herelle. Many had observed that bacteria could suddenly start producing phages without any apparent evidence of infection with a phage, which to some suggested that phages were something produced spontaneously by bacteria itself, not due to an infection. Plus, Bordet had witnessed phages suddenly having no effect on the very strain they were meant to kill.

It is now known that phages can persist quietly within bacterial cells for many generations, their DNA lurking quietly and being copied each time the bacteria replicates itself. The lurking phage DNA, known as a prophage, turns off any of its genes that could harm its host, and sometimes even switches on genes that help boost the long-term prospects of their landlords – for example, blocking other phages entering the cell or providing genes to boost its metabolism. This less violent lifestyle can be just as effective a way to replicate as forcing a cell to make hundreds of new virions and bursting it open. But the lurking phage DNA is a volatile guest – it always has the potential to switch to the more violent virus-producing mode whenever it wants, for example when it senses the host has plentiful resources to make lots of viruses, or if its host is struggling and the phage wants to escape out into the world and find another place to live and replicate.

The full detail of the radically different paths phages may take when infecting a bacterium was not fully understood until the 1950s. And so, for many decades, scientists watched as phages infected bacteria but stopped producing new viruses, or bacteria, seemingly uninfected by phages, suddenly started to produce phages. No wonder people were confused by the nature of phages, and no wonder phage therapy didn't always work.

Just when it seemed that d'Herelle's rise to prominence was unstoppable, Bordet and the Belgian group came across an old paper in *The Lancet* that would mire him in controversy for the rest of his scientific career.

The details were familiar: an ultra-microscopic, filterable substance; plaques of bacterial death on a plate; strange powers of growth and replication. But the author was an Englishman: Frederick Twort. And the paper had been published in 1915, two years before d'Herelle presented his work to the academy. To the Belgian group, it was gold dust: not only was the despicable and distrusted d'Herelle wrong – he was a plagiarist!

Now the gloves really came off. A long and tedious dispute followed, as d'Herelle desperately tried to prove his and Twort's observations were different. For one protégé of Bordet, André

Gratia, proving that the two phenomena were identical became the main focus of his career.[61] It was Gratia who had discovered Twort's largely unnoticed paper from *The Lancet*, by sheer fluke: he had been recommended to find a paper in a volume of the journal from 1915 but was not given a clear reference, and was skimming through an entire year's worth of articles looking for it when he came across Twort's description of a phage-like substance.[62]

The Belgian group continued to promote the work of scientists who disputed the idea that phages were viruses, while actively hampering d'Herelle's responses. Many letters between the senior figures at the Pasteur Institute and Gratia reveal a shocking network of self-interest and treachery towards their colleague.[63] Senior figures such as Roux, Calmette and Bordet strategised about the most effective ways to damage d'Herelle's credibility, with the younger and less-well-known Gratia doing the scientific dirty work and reporting back to them. Twort, not having worked on the subject for many years, embraced the sudden interest in his work, and after considering the alternative theories on the nature of what he'd seen supported the establishment theory that bacteriophage was an enzyme.

Twort and Gratia became good friends, with Gratia pushing for Twort to be credited not just with the discovery of the bacteriophage phenomenon, but with the invention of phage therapy, too – a field in which Twort had done precisely no work at all. Gratia claimed d'Herelle had simply taken Twort's hard work and popularised it with a catchy name. (A hundred years later, one might argue that the name 'bacteriophage' actually has made bacterial viruses seem obscure compared to all other viruses, which are just called, well, viruses.)

Scientists opposed to d'Herelle's ideas referred to the subject as the 'Twort–d'Herelle phenomenon', reinforcing the idea that d'Herelle was not the first to observe it and presumably to avoid using the 'catchy' term bacteriophage. The network dedicated to beating d'Herelle down was strengthened by the exchange of favours – from publication in prestigious journals to research opportunities. In one letter, Gratia stressed to Calmette the importance of publishing their latest paper on Twort's observations before the Nobel Prize committee made their decision that year.[64] He had heard rumours that d'Herelle's work on phages and immunity might scupper the chances of Calmette getting the famous prize for his BCG vaccine. It is believed d'Herelle was nominated for a Nobel Prize at least twenty-eight times,[65] possibly more, but the campaign to sow confusion and distrust in his work ensured he never won it.

The dispute rumbled on and on and remains unsettled: the famous mid-century phage scientist Gunther Stent believed d'Herelle knew about Twort's work and lied about it, while the French science historian Alain Dublanchet suggested in 2007 that a letter from d'Herelle to Twort, which supposedly proves he knew about his work, may be a forgery.[66] Today, Twort is generally recognised as having 'priority' in the observation of the action of phages on bacteria – priority being important science jargon for 'who discovered it first'. But d'Herelle is recognised as the founding father of phage science and phage therapy.

In the summer of 1927, a decade on from that first presentation on bacteriophages, d'Herelle's claims about the effectiveness of phages against bacterial disease were put to the test in the largest and most expensive field trial of phage therapy yet, known as the Bacteriophage Inquiry. This involved distributing thousands of ampules of phages to local dispensary doctors in rural villages in India, who were ordered to give them out at the first sign of cholera. Two industrial-scale manufacturing centres were set up in the Punjab region to create the anti-cholera phage mixtures,[67] and eventually, the studies involved over a million people.

When d'Herelle handed over the field trials to the 'trusted hands' of Yugoslavian microbiologist Igor Asheshov, phages were simply dumped in wells. These wells were not only where the villagers drew their water from but were also sources of water for thousands of pilgrims during summer pilgrimages. The incidence of cholera that summer in the region was certainly far lower than normal – there were about eight times fewer cases among the pilgrims compared to people in the wider region. D'Herelle would later claim that figures from the local health authority showed that the cost of his treatment came to only 25 rupees per person, while previous campaigns involving mass vaccination and disinfection of water supplies was more like 3,000.

But seemingly positive results and excitable endorsements from local health officials were undermined by the vastly over-ambitious scale and slapdash scientific methods used by d'Herelle, Asheshov and their sponsors. What's more, the trials were conducted at a time of increasing political turmoil and unrest in India. Mahatma Gandhi was encouraging non-compliance and civil disobedience against the British authorities, and many local 'headmen' or local dispensing doctors did not follow the rules of the vast trial or keep the records that they had been asked to. Such was the scale of the distribution of phages, it was also not always clear who had received doses or whether those who had had been diagnosed with cholera correctly.

Ironically, what was fatal for the credibility of d'Herelle's giant experiments in India was the booming popularity of phage therapy itself. The exciting new treatment was in such demand, with doses being traded and transported all over the region, that many doses ended up in villages that were meant to be control groups in the field trials. By the late 1920s, d'Herelle's Laboratoire du Bactériophage in Paris was producing over 100,000 doses of cholera phages a year, much of it being shipped to India, on top of the large manufacturing centres set up for the mass experiment.

By 1934, that figure had risen to over 800,000 doses a year.[68] As well as d'Herelle's private lab, other pharmaceutical companies in the UK and Germany were producing lines of bacteriophage products too.

The scientific committee from the Bacteriophage Inquiry remarked:

... This measure has become so popular in recent years that a very considerable quantity is used in every district in Assam during the cholera season. This can only be described as disastrous from the point of view of the experiment.[69]

In other words, there was no way of comparing how well the phage treatment compared to places without phages, because d'Herelle's phages were being used everywhere. With the cholera epidemics killing roughly 60% of patients who did not have access to phages, it became increasingly clear that it would be unethical to prevent thousands of people accessing this treatment just to keep the experiment on track.* The whole thing was eventually called off.

Adding to the confusion, widespread reforms across India had improved sanitation and therefore reduced the overall risk of cholera in many areas during the course of the study. The committee's remarks on whether phages should be recommended for widespread use in India were damning:

> It is practically impossible to obtain reliable figures for comparable treated and untreated cases under field conditions . . . There is so far practically no conclusive evidence from Assam or elsewhere on the most important question of all, namely, should epidemics of cholera be treated by bacteriophage rather than by the accepted methods of disinfection and vaccination.

The first large-scale trial of d'Herelle's 'miracle-cure' bacteriophage had been a disaster. The study ended with no reliable or comparable data, a problem that would become a hallmark of studies of phage therapy all through the twentieth century; it remains an issue to this day. The endless bickering and confusion about what phages even were was also affecting scientists' perceptions of phages as a medical treatment, too. The esteemed journal *Science* stated in 1929 that phage therapy 'has fallen short of fulfilling this promise

---

* This interesting ethical dilemma was explored in the Pulitzer Prize-winning novel *Arrowsmith*, by Sinclair Lewis. The main character, said to be based on d'Herelle, changes his mind about whether to treat a large group of patients in a study after his wife dies of plague.

because the men who had to use it have not understood it well enough'.[70]

Almost a century later, we can study phages with a precision and level of detail greater than d'Herelle could ever have hoped for. We can watch them in action with powerful microscopes, study the individual proteins and genes within them that drive their behaviour, and even engineer them to have entirely new characteristics and features. We can use artificial intelligence to crunch through unfathomable quantities of data and predict how hypothetical combinations of phages and bacterial species will interact. Though there is much that is still unknown, we understand some of the intricacies of phages that caused so much confusion about their physical nature, and why their use in medicine could be near-miraculous in one patient and completely hopeless in the next.

While d'Herelle understood that the right phages for the job could be found living where the bacteria of interest were also found – that is, the poop of the patients he was treating – the rules around which phages can infect which bacteria were not well understood. We now know that some phages are generalists – weakly infectious against a broad range of differ- ent species of bacteria – while others are more specific, finely tuned to infect and evade the defences of one particular bacte- rial species extremely efficiently. Many phages are in fact hyper-specific, targeting only particular strains of bacteria, that is, a very particular subspecies of the bacteria with slightly different characteristics to the others.

Thanks to developments in the science of molecular biology,

we know that phages have highly specific receptor molecules on their tail fibres (or, for tail-less phages, on the tips of spike proteins sticking out from their outer coat), which fit like a lock and key on certain molecules displayed on the outer surface of bacteria. Different bacterial species are covered in a huge range of different proteins, sugars and other molecules, and some have appendages such as flagella (whip-like tails that help propel them around), fimbriae (hair-like protrusions that help them attach to surfaces) or pili (fine tube-like structures that help them connect to other bacteria). It is these surface molecules or structures that phages must sense, recognise and bind to if they are to infect the right host. (One phage of *Salmonella* bacteria, nicknamed 'cowboy', uses its long and flexible tail to lasso the flagellum of its host before pulling itself down it to infect the cell itself.) As bacteria constantly evolve to evade phage attacks, the fine structure of these potential binding sites change, too. Phages therefore constantly adapt their receptor sites in response. Over billions of years this has led to a vast number of very slightly different strains of both bacteria and phages, all locked in a never-ending dance with the other shifting microbes around them.

Some species of phage are able to shuffle the molecules that make up the receptors on their tail fibres, effectively throwing the dice to see if the new configuration helps them find a match. And some bacteria are able to remove entire surface membranes or dump their tail and hair-like appendages to prevent phages having anything to cling on to. In other words, bacteria can quickly become resistant to the phages that are attacking them, and phages can change the way they attack bacteria in response.

These tiny war games happen constantly even in small populations of microbes over short periods of time. The constant adjustment of bacterial defences and subsequent adaptation of phages means no one bacterial strain can ever become too dominant, and no one phage can wipe out an entire bacterial population. This creates not just diversity but also a kind of stability in microbial ecosystems.

The first decades of phage therapy were conducted with a poor understanding of the lifecycles and host range of phages, and the ways bacteria can adapt to evade them. By the 1940s, labs and pharmaceutical companies in Europe and America had begun selling dubious phage products, from small operations like 'the German Bacteriophage Society' to pharmaceutical giants like Abbott, Eli Lilly and Squibb & Sons (now known as Bristol-Meyers Squibb), many of which misunderstood or wilfully ignored the complex principles of phage biology that d'Herelle was working so hard to understand.

Finding a mix of phages that could cover most locally circulating strains of a certain disease was one of the greatest challenges of using phages, as d'Herelle had found several times. When he travelled to Egypt to help treat outbreaks of bubonic plague, caused by the bacteria *Yersinia pestis*, he found that the phages he had used to defeat the disease in the Far East were completely ineffective against local strains of the same bacteria in Cairo.

Few people, if anyone, understood how to purify and concentrate mixtures of phages that could work for a broad range of different bacteria. Yet one advert for the dried phage product 'Enterofagos', produced in Germany, claimed it could treat 'typhoid fever, paratyphoid fever, dysentery, colitis, all types of diarrhoea . . .' Another product of the same name sold in London went further, claiming it could quite impossibly treat hives (an allergic reaction), eczema (an allergic and genetic condition) and herpes (caused by a virus, not bacteria).[71] By the late 1920s and early 1930s, injudicious use of any old phages to treat any old infection – with predictably poor results – meant their reputation as a wonder drug was starting to fade in most countries. But not all.

# PART 2

*Forgotten Phages*

# 4

## Stalin's medicine

There are a few places in the world today where you can walk into a pharmacy and buy a packet of smartly branded medical-grade bacteriophages. After a brief chat with a pharmacist, you swig a few millilitres of yellowish liquid from a small glass ampule or rub a phage-infused ointment onto your spots or abscess or wounds, and hope the trillions of viruses you have just ingested or absorbed can counter whatever bacterial malady is plaguing you.

One of these places is Georgia, the mountainous former Soviet state squeezed between Russia's southernmost tip and Turkey. In Georgia's capital, Tbilisi, crumbling Soviet-era tower blocks rise up between grand old wood-fronted apartments, ancient churches, mosques and fortresses. The concrete-coloured river Mtkvari runs fast and high through the city's discombobulating mix of Georgian, European, Soviet and Middle Eastern architecture. Road signs, shop fronts and billboards are all embellished with the unique and ancient looping lettering of the Georgian alphabet, a script unlike any other in the world.

Every year hundreds of foreign patients make the difficult journey to Tbilisi seeking a therapy the rest of the world abandoned long ago. They often arrive in a miserable state having endured long and unsuccessful courses of antibiotics for their medical problems – from urinary tract infections to purulent surgery wounds, chronic respiratory problems or infected burns. Set back from a busy, winding road on the steep western bank of the Mtkvari is the Eliava Institute, the first and still one of the only centres in the world dedicated to phage research, production and therapy. Scientists and doctors first started exploring phages as a way to treat bacterial infections here almost a hundred years ago, but unlike the rest of the world, they never stopped.

Two of the six commercially available phage products manufactured and sold here, Pyo-bacteriophage and Intesti-bacteriophage, are essentially the same preparations produced by Felix d'Herelle's Laboratoire du Bacteriophage in the 1920s,[1] albeit with a regularly updated mix of phages more relevant to the common strains of bacteria circulating today. The main building of the institute, built in the 1920s, remains spectacular, its towering Doric columns topped with the flags of Georgia, the United States and the European Union, which have helped fund research here over the years. But inside is sparsely furnished and, when I visit, eerily quiet.

In the large and slightly ramshackle grounds, there are newer buildings – a pharmacy, a diagnostic centre, and an outpatient clinic. There's also some things you might not be used to seeing at a treatment centre, like the derelict shack subsumed by rust and vegetation among the trees. In the institute's heyday, a large hostel across the road helped accommodate many hundreds of staff, but it has since been converted to residential apartments, with basketball courts and washing strung out on the balconies. And a large cottage that was once meant to house the king of phage therapy, Felix d'Herelle, and his family now sits behind huge spike-topped walls, adorned with multiple CCTV cameras and microphones, having been commandeered by the KGB in the 1990s. It is now under even more mysterious ownership.

Dr Nina Chanishvili, the institute's head of research and development, takes me up the main building's imposing stone stairs, through a maze of barren corridors, and up a flight of small wooden steps to a warren of offices, labs and a homely little kitchen, full of boozy gifts from patients and journalists around the world. Over a generous lunch of tea, bread, aubergine spread, cured meat, cheese, cake and Georgian pastries, Chanishvili recounts the remarkable and at times upsetting history of this place, and how the idea of phage therapy has endured here like nowhere else.

She shows me a collection of wonderful black and white images of former staff with Felix d'Herelle and their families, including evocative holiday snaps taken by the great man while holidaying here in Georgia and in Russia. What I assume is an antique wooden cupboard is actually a 1920s electric incubator for warming cultures of bacteria, bought to Georgia from Paris by d'Herelle. It still works.

The original plan for this place was to build a world centre for phage research, spread over a 17-hectare campus – with a 600-bed hospital to treat patients, industrial phage medicine production facilities and, of course, the on-site accommodation for d'Herelle. Two small details on the building's blueprints hint at the turbulent history that almost destroyed this place, as well as the heroes that kept it running and the idea of phage therapy alive. On the first page, the name of the institute's founder and first director has been blotted out with ink. But on the same page, in tiny script and in pencil, someone has added the name back in: George Eliava.

'Somebody wanted to ensure his name was not forgotten,' says Chanishvili, a small and often stony-faced professor with a thick fringe and elaborately framed spectacles. 'But you can see, even when that period was over, they were still afraid. It's small and in pencil so it can be erased quickly if needed.'

As Felix d'Herelle fought off attacks from the scientific establishment in France, he found an unlikely ally in George Eliava, a young Georgian scientist who had come to Paris to learn from the world's best microbiologists. They made an odd pair – Eliava was almost twenty years younger and a charismatic, sporty socialite, known for his charm, gregariousness and funny pranks. His wife was a famous soprano and soloist, and Eliava was friends with a huge circle of Georgia's most celebrated poets, artists, musicians and engineers. He held lavish parties and raced horses.

Early in his career, Eliava had seen a flask full of cholera bacteria, isolated and cultured from Tbilisi's river Mtkvari, seemingly vanish over the course of a meeting – a phenomenon he could not explain until he came across the work of d'Herelle. After studying in Geneva and Moscow, Eliava had become interested in bacteriology. (The influence of a powerful aristocratic aunt had saved his career after he was banned from attending university by the government for joining revolutionary protests.)[2] In a remarkably steep career trajectory, by his early twenties, Eliava was head of a bacteriology laboratory in Trabzon, an outpost of the Russian empire, now part of Turkey. And when d'Herelle published his infamous paper announcing the discovery of the phage in 1917, the twenty-five-year-old Eliava was head of the Bacteriology Laboratory in Tbilisi.

In the chaos of the Russian Revolution that same year, Georgia had claimed independence from Russia, and many Georgians travelled to Europe in the hope of bringing new knowledge and ideas back to their newly independent country. In 1919, Eliava went to the Pasteur Institute to learn the latest techniques in microbiology and vaccinology and was soon performing experiments that supported d'Herelle's ideas on the viral nature of phage, which endeared him to the notoriously spiky biologist.

According to Eliava's granddaughter, the embattled d'Herelle was once so pleased with the results of Eliava's work that he rushed back to Paris from the French countryside and burst into the Pasteur Institute's entrance, asking 'Where is this Eliava? Show him to me!' When the round-faced, cheeky-looking Eliava emerged, the odd couple apparently embraced, kissed and from then on were 'like father and son'.[3] Clearly convinced d'Herelle was onto something, Eliava continued to visit the Pasteur Institute throughout the 1920s and 1930s to study phages, making several important contributions to early phage science.[*]

Back in Georgia, Eliava rose to the position of chairman of the microbiology department at Tbilisi University, and by the late 1920s was beginning to develop grand plans for a dedicated institute for the study and use of bacteriophages. Of course, no

---

[*] Eliava's achievements included the discovery of lysins, the special enzymes phages use to rupture their bacterial host from the inside out, now being explored as the basis for an entirely new class of antibiotic drug.

such institute would be complete without the master himself, Felix d'Herelle, and Eliava offered his pal the run of the place, promising him an honorary position, freedom to conduct whatever phage research he wanted, plus a beautiful on-site cottage for his family. The stage was set for phages and phage therapy to finally get the scientific attention they deserved.

There was just one problem: by this point Georgia had been brutally retaken by the Russians. And at the head of the latest iteration of the Russian super-state was another Georgian – Ioseb Besarionis dze Jughashvili, better known in the West as Joseph Stalin.

Despite the controversy that seemed to follow d'Herelle about, and his war of words with colleagues in Paris, the founding father of phage science had continued to attract interest and admiration across the world – at least from those who'd never crossed him. While perceptions of his theories and of phage therapy fluctuated, he had maintained a high profile and a knack for commanding the attention of journalists, health officials and foreign dignitaries on his travels. The Royal Academy of Sciences of Amsterdam had awarded d'Herelle the Leeuwenhoek Medal for his discovery of bacteriophages, an honour given just once every ten years to people who have made distinguished contributions to microbiology.\* The only

\* The award is named after Antonie van Leeuwenhoek, the Dutch lens-maker and self-taught scientist, who, in the 1670s, first made a looking glass powerful enough to reveal that the world is awash with tiny living microbes (or 'animalcules' as he called them).

other French scientist to be so honoured was d'Herelle's hero, the great Louis Pasteur.* *The New York Times* had covered d'Herelle's work again in 1925, under the wonderful headline 'Tiny and Deadly Bacillus Has Enemies Still Smaller', and d'Herelle's story even inspired a Pulitzer Prize-winning novel, Sinclair Lewis's 1926 *Arrowsmith*, which went on to be a huge hit, adapted for stage, radio and film.

In 1928 he was offered the prestigious position of chair of bacteriology at Yale University. A press release from the university proudly announced: 'Famous scientist appointed to faculty of the Yale School of Medicine',[4] and the budget of its entire medical school was adjusted to meet the needs of its new superstar. But ever his own worst enemy, d'Herelle was soon irritating his new colleagues. He left the hallowed grounds of the university only a few days after arriving to continue a long and lucrative lecture tour of the United States, and regularly returned to France to oversee the production of therapeutic products at his phage lab in Paris.

Atop his reputation for absence, some of his colleagues began to see him as lazy and a hypochondriac.[5] This may be a bit harsh, given that d'Herelle had picked up a range of nasty bugs over the course of his life as a microbe hunter – amoebiasis (twice), malaria (several times), an unknown fever in Egypt and breathing problems caused by a botched attempt to inoculate himself with the bacteria that cause tetanus, *Clostridium tetani* – which probably explains why he was subsequently

* Like d'Herelle, Pasteur had also begun his career working on fermentations, then moved onto pest control, before moving into fighting disease – so d'Herelle's perhaps 'life plan' to emulate this scientific icon's career was really not so laughable after all.

plagued by poor health for most of his life. D'Herelle had also negotiated a salary of $10,000 a year, which equates to hundreds of thousands of dollars today and was a frankly ridiculous amount of money for the time – yet he began to complain that he was not being treated fairly and demanded more money to stay at Yale all year round. His request might seem improper at the best of times, let alone in a country falling into the Great Depression. He was burning his bridges, yet again.

Having been bullied out of what was supposedly one of the most prestigious scientific institutes in the world in Paris, and in constant disputes with his employers at Yale, by the 1930s d'Herelle was becoming disillusioned with Western science and a medical profession he saw as corrupted by profit. With the expanding Soviet regime keen to recruit high-profile defectors from the West, he received offers to start a bacterio-phage institute in Moscow, as well as the invitation from Eliava in Tbilisi. While d'Herelle had toured much of Russia, he preferred Georgia, known for its beautiful scenery and food, hospitable people and a mild climate which would soothe his respiratory ailments. And, of course, there would be the support of his good friend 'Giorgi'.

As he flirted with a move and took trips around the Soviet Union with his wife, the famed newspaper *Pravda*, by then the official newspaper of the Communist Party, described him as 'one of the most outstanding microbiologists in Western Europe'.[6]

In his miserable time at Yale, d'Herelle had written despairingly of the destitution the Great Depression had inflicted on America, when people were unable to live even 'a basic existence' despite the country hitting an 'apex of wealth' just a few years previously. Plus, he had seen how phage therapy – requiring complex administration and infrastructure but not necessarily generating profitable products – had not been a good fit within the companies chasing profits and doctors looking for quick fixes. Manufacturers had ignored his carefully established rules on how to use phages in medicine, making clumsy mixtures that they could flog to the masses.

In contrast, the Soviets were experimenting with a new, staunchly anti-capitalist version of society. Despite plenty of signs that the USSR under Stalin was no utopia, looking and thinking like a communist was still hugely fashionable among radicals and intellectuals in many parts of the world. Plus, in order to fight infectious disease, the Stalinist government was building a vast state-funded healthcare system which included networks of microbiological surveillance labs in different parts of the empire. D'Herelle's family maintain that he wasn't committed to any political ideology, and was certainly not a Stalin sympathiser, but he clearly believed the Soviets were developing the kind of infrastructure that was needed to handle the complexities of treating patients with phages.

In his third book on bacteriophages,[7] published in 1935, d'Herelle dedicated the Russian translation to Stalin – a standard requirement for anything being published in the USSR at the time – but in his own inimitable style, he conveyed his hopes for the Soviet experiment by taking a swipe at the entire Western establishment:

I have written [this book] for the scientists of the USSR, this wonderful country which, for the first time in history, did not choose irrational mysticism as its guide, but sober science.[8]

It may have been the relentless attacks from his French peers, the damning reviews of phage therapy in major journals, or simply the irresistible draw of the fame and glory he felt he deserved; whatever it was, d'Herelle decided to turn his back on Paris and Yale, Europe and America, offering himself to what would become one of the most brutal, paranoid and murderous regimes in history – taking his unique understanding of bacterial viruses with him.

In 1930s Georgia, people were seen as either 'old' or 'new' communists. Roughly speaking, the older communists were loyal to the original revolutionary ideals of Lenin and Trotsky, were often intellectuals or members of the celebrity elite, and were critical of the new and stifling Soviet bureaucracy. The 'new' communists, loyal to the increasingly tyrannical Stalin, included local state bureaucrats, informers and spies who suppressed individualism in pursuit of an orderly and supposedly productive populace. They saw political critics, non-Soviet nationals, the bourgeoisie and even land-owning peasants as threats to the success of communism and the Stalin regime.

The flamboyant Eliava – well-educated and well-travelled, with an aristocratic background, was certainly of the old class. At a time when almost any display of decadence or celebration of foreign ideas could be seen as hostility to the Spartan ideals of the regime, Eliava's reputation for racing horses and hosting lavish parties would seem to make him a marked man. That was before he started picking fights with one of Stalin's most powerful and sadistic allies, the infamous Lavrentiy Beria.

Of the new class of communists, Beria had begun his career in 'state security', before becoming a local Georgian politician. He rose up the ranks of the Soviet system to become secretary of the Communist Party in Georgia, head of the Interior Ministry of the Soviet Union and eventually, deputy to Stalin himself. Beria's various grandiose titles disguise his real role in Stalin's murderous Soviet administrations: he was effectively either working for or in charge of the secret police that repressed, persecuted, purged and executed millions of Soviet citizens for decades before Stalin's death in 1952. This small, bald psychopath with tiny round glasses organised killings that ranged from ethnic cleansing and political purges to depraved sexual attacks from his limousine late at night.

Eliava was not one to be cowed by such figures. As part of a rebellious parlour trick, he had trained his dog to shake its head and refuse a treat at the word 'take', which when repeated in Russian sounds like 'Beria'.[9] In Georgian society, it became an open secret that Eliava and his friends liked to defy and mock Stalin's cronies, especially the wicked Beria. 'He would say exactly what he thought, openly criticising the government,' Eliava's great grandson Dimitri Devdariani tells me. 'In the 1930s Soviet Union you just did not do that. He was by nature a free spirit, and nothing could contain that freedom.'

Chanishvili, who has studied Eliava's life extensively and interviewed one of the last people alive to have worked with him, has a different take on this apparent brazenness. 'He acted that way because everyone in his circles acted that way.'

By 1936, Eliava had used his charm and his contacts to secure approval and funding for the extraordinary institute dedicated to phage science. It was to be built on a vast plot of sloping land west of the Mtkvari, the river in which he had unwittingly found cholera-clearing phages decades earlier. The building's grand entrance would be framed by an imposing set of huge columns, and the building's low, wide wings would flare out in a modern, asymmetrical layout. With a budget of 13 million roubles, the enormous complex was to have clinics and a teaching hospital for phage therapy, research labs for basic phage research, hostels for staff – and even stables for horses to provide the blood-based serum required in the labs. The rooms were to be furnished with the latest scientific equipment, and d'Herelle and his family were promised their own chauffeur in addition to their charming on-site accommodation.

But things were far from rosy in this beautiful part of the Soviet Union. Stalin was by now using forced labour across his empire's vast territories and had begun to purge his political and cultural enemies from public life. Statues and portraits in honour of the dictator became a more prominent feature of everyday life, as dissent and political opposition slipped into the shadows. Eliava himself had already been arrested as part of a campaign of harassment against leading figures in science and agriculture by party loyalists acting outside Georgia's normal judicial system.[10]

The Soviet Union had also been pushing its academics down a self-defeating ideological blind alley known as 'Lysenkoism', which threatened the careers and lives of many scientists, but especially biologists. Trofim Lysenko was a Russian biologist who had made a series of claims about the possibilities of crop breeding that were at odds with Western theories of genetics at the time, yet appealed to a Soviet regime battling vast famine. His ideas were based on the discredited theory that organisms'

86

genes can change over their lifetime depending on their experiences and environment, and that these changes can be passed onto future generations – which is a little like saying that a thin man who develops enormous thighs from cycling everywhere will have a son or daughter with enormous thighs, regardless of whether they cycle or not.*

This was markedly different to Darwin's central idea that the genetic make-up of organisms changes gradually over time as the most successful combinations of genes survive and reproduce. Lysenko's assurances that his theory could improve crop yields were so important to the Soviet regime that they promoted him as a genius who could save Soviet farming, despite his ideas being largely dismissed in the West. Lysenko then used his platform to denounce the Western geneticists who threatened his theories, calling them 'fly-lovers and people-haters'. The Soviet propaganda machine continued to celebrate his work, denounce his critics and cover up his failures for so long that they eventually had to ban conventional genetics research across the Soviet Union to stop people proving it was all rubbish. The Stalin regime was, in fact, so committed to maintaining the idea that Lysenko was a Soviet genius – even as crop yields fell – that thousands of biologists were sacked, sent to labour camps or executed over the following decades.

---

* In recent years, the scientific field known as 'epigenetics' has explored how different genes in different cells in an organism can be expressed differently at different times. There is some evidence that these modifications to our genome are dependent on our environment and our life experiences, and even that some of these changes appear to be sometimes passed on to different generations. However, the effect is subtle, and organisms certainly cannot change which genes they have in their lifetimes.

Despite these oppressive conditions, Eliava continued to live with a breath-taking and some might say suicidal fearlessness. It is said that one day, when Beria was suffering from a mysterious illness, Eliava – being the foremost expert on infectious disease in Georgia – was summoned to take a blood sample from him. The story goes that when Beria joked weakly to Eliava not to take too much blood, Eliava replied sharply that Beria was sucking the blood of the nation. Another version of the many stories about the rivalry was that the miserable-looking Beria – who by 1936 was the leader of the Soviet Union in Georgia – was intensely jealous of Eliava, whom he had known since childhood. Beria was apparently besotted with a beautiful woman who was married to a friend of Eliava, and after having her husband 'disappeared' by the secret police, was dismayed to see her walking down the street with Eliava. This was the final straw for the insecure, brutal administrator, according to some.* Other accounts suggest that Eliava had offended Beria by bypassing him and appealing directly to Stalin and the ministry of health for funding for the enormous phage institute at Tbilisi.

All that is clear is that by 1937, just as the main buildings of Eliava's enormous institute were being finalised, he was suddenly arrested in front of his family while hosting a dinner party. 'Look after your mother,' he apparently told his stepdaughter, Hanna, not knowing that his wife would soon disappear too. Even Hanna would eventually be arrested and sent to a forced labour camp.

---

* Spicier versions of the story suggest that Eliava was not protecting his recently widowed friend, but dating her, despite being married himself.

In one photo of Eliava, taken just two weeks before his disappearance, Eliava sits unsmiling in the middle of a group hug as a group of joyful female colleagues celebrate a birthday. The haunting image is now on the wall of a small museum in the Eliava Institute, and even Chanishvili, not one for sentimentality or innuendo, says she sees a sense of foreboding in his eyes. On the other side of the room is a photo of workers at the institute in 1937, with a ghostly silhouette where someone has been crudely erased from the picture.

Georgia's main newspaper, *Kommunisti*, reported the state's claims that Eliava had been preparing dangerous bacteria and poisoning wells 'to kill the Soviet people' on the orders of enemies of the USSR. 'This animal got what it deserved', the rabid article continued, providing no details about his imprisonment or trial. As with most forms of misinformation, the story was distorted from a kernel of truth: Eliava did, of course, work with dangerous bacteria, and had probably been studying the public health benefits of adding phages to wells of drinking water.

The same fate would eventually meet almost all of Eliava's famous and intellectual friends,* and hundreds of thousands of others, in a spasm of paranoid violence that year that became known as the Terror, or Great Purge.

'We still do not know how or when exactly they perished,

---

* Chanishvili has found details of an incident involving a party attended by Eliava and his friends which had annoyed Beria, who lived opposite. When Beria sent a subordinate to summon the men to his apartment, almost all of them laughed and swore at the request, bar one person, who went over for a dressing down. Every person at the party was eventually arrested and executed, except the one person who complied with Beria's orders.

although we can assume,' says Eliava's great grandson, Devdariani. 'The KGB and secret services did their damnedest to erase his memory – executing his wife and even arresting his stepdaughter, my grandmother. Their photos were confiscated, documents, all his scientific works destroyed. The day he was arrested he was due to be presenting work at a major conference in Moscow. His article was destroyed, and I'm sure there were other works that never saw the light of day.'

The development of phage therapy into a reliable treatment now faced far more sinister problems than the trifling academic disputes that had dominated the previous decades. D'Herelle was fortunate that at the time Eliava and his wife were arrested, he was in Paris, awaiting a visa to move to Georgia. He had intended to settle permanently in Tbilisi, and for a time, continued to write to his friend, seemingly unaware of what had happened.

Heartbroken by the death of his friend, and disappointed that his hopes for a new scientific utopia had been so brutally naive, d'Herelle never wrote anything of his time in the Soviet Union or Georgia from then on. While his Laboratoire du Bacteriophage continued to churn out phages in Paris, his 'missionary zeal to proselytise for phage therapy' began to wane, according to his biographer William Summers.

He became embroiled in a dispute with the investors of his lab over some advertising he did not agree with. Eventually, it went to France's Supreme Court, and d'Herelle, with terrible timing, decided to move from Paris to Vichy just before France fell to the Germans and the town became the seat of the collaborationist government. With his Canadian citizenship, he was kept under virtual house arrest, and his enforced inactivity effectively ended his research career* and exacerbated his health problems. He stuck rigidly to his belief that phages were the human body's main defence against bacteria, despite the growing body of work showing that a complex system of white blood cells and antibodies were involved, for the rest of his life. He died of pancreatic cancer in 1949.

Despite d'Herelle's remarkable breakthroughs on the nature of viruses, his name soon faded, and his work is largely unknown and uncelebrated outside of the virology community. (Frederick Twort died a year later, in 1950.)

Eliava, meanwhile, who had brought phage science to the Soviet Union, was removed from the empire's history with a bullet, a shovel and the stroke of a bureaucrat's pen. Even his family were unaware of much of his scientific work until recently. But this bloody episode was not the end of phage therapy in Georgia. Just like a phage dropped into a flask of the right host, the essence of this unusual idea began to grow and grow and grow.

---

* Although he had no paid research position, he did continue to come up with ideas and experiments, such as finding a phage to prevent flatulence. 'Test after eating cabbage,' he wrote in a short note to himself.

# 5

## Phages versus the Nazis

Before Stalin's rise to power, news of the discovery of bacterial viruses had attracted widespread interest across Russia and the newly founded Soviet Union. Life in many areas of this vast empire was dominated by outbreaks of disease, which found fertile ground in an ill-nourished population exhausted by civil war – sometimes compounded by shortages of soap and cleaning products.[11]

Lenin, installed as the head of the government following the 1917 revolution, had made fighting infectious disease a priority; the scale of the outbreaks spreading through the population was so alarming it threatened his grand plans to rebuild the country. Along with the age-old epidemics of dysentery and cholera, the young Bolshevik government had rampant typhus and smallpox epidemics to deal with, on top of the influenza pandemic of 1918.

Even by official estimates – which given the post-revolution chaos would likely have undercounted cases of disease – a quarter of all Russians had typhus in 1919. That's 30 million or so people. Three million died from it,[12] and the Russian population shrank each year for the next nine years.[13] In 1919 Lenin proclaimed that the country should effectively throw all of the resolve and resources away from civil war towards fighting disease. Referring to typhus, caused by the bacterium *Rickettsia prowazekii* and spread by lice, he commented: 'Either the lice defeat socialism, or socialism defeats the lice.'[14]

By the 1920s, the Soviets had created the beginnings of a state-run, socialised healthcare system, and a network of bacteriological surveillance labs was set up to track disease across its vast expanses of territory. Just as in the rest of the world, reliable treatments for bacterial diseases were few and far between. Inspired by d'Herelle's development of therapies against dysentery, typhoid, plague and cholera, trials using phages both to treat and prevent bacterial disease had begun as early as 1929 in what was then Soviet Ukraine.[15] Rather than upsetting people, d'Herelle's thrilling theories about phages' role in outbreaks and resolution of diseases struck a chord with the culture and trends in Soviet science,[16] where microbiologists were interested in the interaction and symbiosis between germs, people and their environment.

Ilya Mechnikov, a titan of early twentieth-century Russian science, had theorised that mammalian immune cells may once have been free-living single-celled organisms that had at some point been recruited by ancestral animals to help us fend off infections. When d'Herelle suggested that phages may also have been recruited by the human immune system to defend against bacterial disease, it stimulated great interest in Russia and the Soviet states.[17]

Despite Eliava's disappearance, and Stalin's claims that his work was deadly treason, the large institute founded in Tbilisi continued to operate in the 1930s and 1940s, albeit under the name of the Institute of Microbiology, Epidemiology and Bacteriophage. Eliava's name and contributions to Soviet science were scrubbed from official documents, and d'Herelle never again set foot in the empire that had once wooed him. But together the men had made significant progress in popularising the idea of phage therapy and research with those who held the purse strings in Georgia and Moscow.[18] Rather than endlessly arguing about phages, the expanding and ambitious empire began to invest in practical ways to make this new way of treating infectious disease work.

Rampant infectious disease remained a problem across the Soviet Union, and with an invasion of Finland being planned, phages were seen as a promising way to help reduce the number of deaths and casualties caused by war.

The Soviets began to explore phage-based treatments not only for the typical war diseases like cholera and dysentery, but also for gas gangrene, a frightful infection of the flesh caused by *Clostridium perfringens* bacteria colonising open wounds. The name comes from the toxic gas that builds up in the discoloured skin around an infected wound – which not only gives off a sweet, putrid smell but creates a horrifying cracking sound when the flesh is pressed. In the days before antibiotics, soldiers maimed by bullets or shrapnel could often only watch as, days later, their limbs began to rot like old fruit in front of their eyes, signalling the urgent need for an amputation.

During the winter war with Finland of 1939, mobile crews of medics tested cocktails of anti-*Clostridium* phages from Tbilisi and Moscow, sending them to the tens of thousands of troops on the front line. When soldiers were shot or injured by artillery, medics poured solutions of anti-*Clostridium* phages straight into the wound, alongside whatever local, general or booze-based anaesthetics were available at the time. Those already suffering from gangrene had their wounds washed in iodine and alcohol before being sprayed with phages. If the infection was severe, phages would be injected into the bloodstream at either the hip or shoulder, with the wounds dressed in phage-soaked bandages afterwards.

The developments were said to have reduced the number of gangrene fatalities by a third, but as with many studies in the USSR at the time, there was no control group to compare the treatment to.[19] What's more, the technology available at the time to purify phages was still far from what we think of as pharmaceutical-grade, and injections of phages would almost certainly contain fragments of the very bacteria that they were meant to be ridding the body of. This could cause serious allergic reactions – or worse, a dose of the very same bacterial toxins that were making the patient ill.

While scientific reviews of phage therapy in the West began to write it off as inconsistent and unpredictable, the Soviets were finessing their military medical procedures and discovering that early treatment was key. The relatively small amount of bacteria found in the tissue at the time of wounding could easily be destroyed if suitable bacteriophages were administered during the first hours, when the bacteria were still on the skin's surface. This early treatment seemed to prevent the formation of weeping wounds and led to faster healing. Less timely and more invasive interventions were more temperamental.

Further large-scale trials of phage therapy took place in Ukraine, Belarus, Turkmenistan and Azerbaijan, as well as cities across Russia, from Moscow to the Far East, and, of course, Georgia, throughout the 1930s and 40s.[20,21] But these studies were not published in widely read Western journals, key information around dosage and side-effects were not recorded and, once again, there was often no control group to help prove that the positive effects really were caused by the phages.

Results weren't always rosy, and opinion on the effectiveness of phage therapy varied even in the USSR. Doctors would often receive batches of phages that did not work, and none other than the Surgeon General of the Red Army, Nikolay Burdenko, reported poor results when using phages to treat *Staphylococcus* infections.[22] In an empire of over 150 million people, spread out over almost nine million square miles, the narrow host range of phages created major problems. Batches of phages produced to treat a certain type of bacteria in one region of the USSR, for example in Moscow or Tbilisi, were unlikely to work on strains of the same bacteria circulating thousands of miles away in the Siberian north, or the central Asian steppes, or Russia's far east. Even the strains circulating in two relatively close cities were different and ever-changing. The production of phage medicines was therefore a never-ending battle to create new phage mixtures that were relevant to different regions at different times.

Yet others saw what phages did to bacteria in a flask and remained optimistic that a refinement of the idea could revolutionise Soviet medicine. The deputy head of the Red Army Sanitary Service, Pëtr Zhuravlëv, declared that if the USSR's scientists and surgeons could make phages more consistent, 'by the time the next wars are imposed on us we will have a powerful medicine with which we will be victorious against all infected wounds'.[23] Zhuravlëv clearly knew what was coming. When the Nazis invaded Russia in 1941, it brought the Red

Army into World War II, and conflict on a scale never seen before.

Infectious disease was so rife in cramped barracks and trenches that even a temperamental and disputed treatment was better than nothing. As the war dragged on and money ran out, some soldiers lacked even guns or uniforms. Phages, crucially, were dirt-cheap to produce. Populations of bacteria could be encouraged to blossom in vast vats of nutrient broth, before even bigger populations of phages exploded forth from those bacteria. As production ramped up in Tbilisi, phage scientists took the perilous journey to the front line to brief military doctors on how to use the viral cocktails.

With phages suddenly flooding both military and civilian populations, the Russian science-fiction writer Boris Strugatsky later recalled how they may have saved his life during the siege of Leningrad:

> Then, in March, I came down with the so-called bloody diar-
> rhea, an infectious disease that is dangerous even for a grown,
> portly man, and I was eight years old, and had dystrophy – a
> certain death, one would think. But our neighbor (who also
> miraculously survived) somehow happened to have a vial of
> bacteriophage, so I lived.[24]

In Tbilisi, phage scientist Zinaida Yermolyeva – who would later be glorified for her work at Stalingrad – took Eliava's pioneering efforts forward, leading further studies on both fundamental phage biology and their use in medicine. She stressed the importance of making phage preparations strictly in accordance with d'Herelle's principles.

Another pioneering female Soviet phage scientist, Magdalina Pokrovskaya, had worked with d'Herelle and was convinced that the phage was a microbe of a microbe – or, at least, a substance 'with creature features', as she put it. By treating the phage like a living, evolving life form, Pokrovskaya developed practices to 'train' specific phage strains with useful character-istics and, by passing them through bacterial stains repeatedly, improve their virulence and efficacy.[25]

Meanwhile in the West, another major scientific review of phage therapy in 1941[26] signalled the beginning of the end for the use of phages in medicine. Just like the major reviews of the 1930s, the analysis – based on English-language studies only – concluded that there was limited evidence that they were effec-tive. Despite the sometimes stunning results that could be achieved with these viruses, official studies remained haphaz-ardly planned and the phages themselves were poorly under-stood – making huge swathes of literature hard to interpret or to believe. To the notoriously monolingual scientific world, the more convincing studies published in Georgian, Russian and even those in French might as well not have existed.

The phage products sold by big pharmaceutical companies were failing too. Unlike the carefully crafted products made by d'Herelle's Laboratoire du Bacteriophage, they were made with little understanding of the complexities of phages and their hosts. Whether they worked for a particular patient was a complete lottery. In fact, it was the dubious and unreliable nature of commercial American phage products that led a group of exasperated doctors to form a group to review medicinal products that became the forebear of the Food and Drug Administration (FDA),[27] which today approves and regulates drugs in the US. D'Herelle considered himself the 'guardian of the method' of phage therapy; he wrote that none of the preparations on the market was 'capable of effecting recovery from infectious disease'.*

On top of this, new chemicals known as 'sulpha drugs', based on chemicals called sulphonamides, were now being marketed as the next big thing in disease control by many pharmaceutical enterprises. However, they did not work for a large group of bacteria including *E.coli*, *Salmonella*, and *Shigella*, and could cause severe side-effects and allergic reactions. They were able to supress infections in a large group of bacteria that included *Staphylococcus* and *Streptococcus*, and soon were being sent to thousands of German and Allied soldiers.

During the war, the Germans also had a steady supply of phages for dysentery – and saw some early successes after testing them on prisoners of war. In one trial, an epidemic of the illness swept through just the prisoners, the unfortunate control group, but not the guards or the prisoners with jobs in the kitchens, who had received doses of phages.[28] But as demand increased, the Germans made phages in vast batches which were not regularly updated to ensure they were active against locally circulating strains. The chiefs of the German Army's medical staff were not satisfied by the erratic results.[29] When the Allied forces confiscated large amounts of the Germans' phages, they tested them on German prisoners of war and were equally unimpressed. There are conflicting reports about the Nazis' interest in d'Herelle as they took Paris: some say the scientist heroically refused to help the Nazis produce phages for their soldiers, others that the Germans felt the phages in his private lab weren't worth taking.

---

* Apart from the products produced by his Laboratoire du Bacteriophage, presumably.

Meanwhile, in labs all around Europe and the US, a major scientific breakthrough was developing that would further dampen the world's enthusiasm for phage therapy: the discovery and development of the world's first 'miracle drug' – a truly effective, reliable, mass-producible and broad-acting antibiotic agent.

# 6

## Parallel universes

'Thanks to Penicillin, he will come home!' exclaimed a patriotic wartime advert from *Life* magazine in 1944, above an illustration of an army medic administering the new drug to a fallen comrade in a faraway land. 'From ordinary mold – the greatest healing agent of this war!'

That year, a network of scientists and pharmaceutical companies had managed to prepare millions of doses of this new antibiotic compound for the war effort, in time for the D-Day landings that summer. 'One sometimes finds what one is not looking for', said the Scottish microbiologist Sir Alexander Fleming, who shared the Nobel Prize in 1945 for his discovery of penicillin. 'I certainly didn't plan to revolutionise all medicine by discovering the world's first antibiotic, or bacteria killer. But I suppose that was exactly what I did.'[*]

As far back as 1928, Fleming had found that a concentrated extract from a mould-like fungus called *Penicillium rubens*[†]

---

[*] D'Herelle had literally found a 'bacteria killer' decades before. Although this quote appears in many articles on Fleming, it may be apocryphal, given that its origin is unclear.

[†] *Penicillium* is a large group of different fungi that have been known to science since the early 1800s, getting their name from the Latin for

seemed to kill or inhibit the growth of a large range of different bacteria. With echoes of Twort and d'Herelle, Fleming had been working with cultures of the common infectious bacteria *Staphylococci* when he noticed that one of his plates had been contaminated – a mould was growing on one of them. The bacteria immediately surrounding the invasive fungus had been destroyed.

The extract, later named penicillin, had the essential properties of what we now call an antibiotic: a compound toxic to bacterial cells but not to human ones. Unlike antiseptics or disinfectants, which are toxic to bacteria but also harmful if ingested or absorbed into the body, antibiotics can be used on or in our bodies to wipe out or suppress the growth of the organisms causing disease without severe side-effects.

It is now known that penicillin works by inhibiting the proper growth of the tough walls that surround bacterial cells, meaning new bacteria cannot be formed and existing ones die. Human cells do not have such walls, and so the compound passes harmlessly through our system (except in the small proportion of people who are allergic to it). Fleming's discovery was extraordinary – the first true antibiotic drug that worked powerfully without severe side-effects.* And crucially, compared to phage therapy, this one drug worked consistently on many different types of bacteria.

---

'painter's brush', in reference to the fine filaments that branch out from these fungi as they consume organic matter.

* Technically, the first drugs to be widely available for treating bacterial disease were the sulphonamides, or 'sulpha drugs', which were sold in the 1930s, but the efficacy of these synthetic chemicals were not as consistent, did not work on as great a range of bacteria as penicillin, and could have considerable side-effects, including rashes, fever and mental confusion.

For a long time, penicillin was considered little more than a scientific curiosity, as Fleming and others working on the compound were unable to extract enough of it to even consider using it in medicine. That only began to change in the 1940s when pathologist Howard Florey and chemist Ernst Chain – who shared the Nobel Prize with Fleming – developed a process to allow penicillin to be extracted in progressively larger quantities. It was agonising stuff: their first-ever patient was an English police constable suffering from horrific facial abscesses after a scratch from a rose thorn became infected – and although his symptoms improved almost immediately after receiving the penicillin, the researchers then ran out of it. The constable's terrible symptoms returned, and he died weeks later. From then on, the tiny batches of penicillin produced were reserved for children, who might require smaller doses. Traces of the drug were even recovered from patients' urine in order to be reused.

Slowly, as wartime governments on both sides of the Atlantic assigned hundreds of scientists to work on the problem, strains of *Penicillium* were found that secreted greater quantities of the drug under the industrial manufacturing conditions developed by Florey and Chain. (One particularly important strain was found growing on a past-its-best cantaloupe at a market in Illinois.) The collaborative efforts helped boost the amount of penicillin scientists could produce in time for the crucial final phases of World War II. And so, the 'antibiotic era' began in earnest, and phages were pushed to the margins – in the West, at least.

In 1943, Franklin D. Roosevelt and Winston Churchill met Joseph Stalin to plot the invasion of Normandy of 1944, and an agreement was made to share the latest Western technological advances with the USSR. Florey and an expert in penicillin production, Gordon Sanders, were part of a delegation of scientists sent to Moscow to share their expertise. Sanders recalled that they took mould samples and 'some of our best purified penicillin extract', effectively gifting this miracle drug to the Soviets. But the British were less forthcoming with the exact protocol for mass production. The Soviets began to develop their own distinct version, known as *krustozin* and, like Fleming, initially struggled to produce large enough quantities to be of use.[30]

And so, phages remained the Soviets' medicine of choice throughout the war. While Western governments invested in penicillin production, at Tbilisi's great centre for phage research, several hundred staff were brewing enormous quantities of phages in floor-to-ceiling copper vats full of host bacteria. Operating round the clock, production was up 500% compared to pre-war levels. The Red Army was now not just dishing out phages to troops and citizens in vials and pills, but scattering them in food, sewage and even into the soil around troop battalions in the hope of warding off infections. It is estimated that more than 200,000 *litres* of bacteriophages for wound infections alone were produced for the Soviet army during the war years.[31]

Although a tenth of the Georgian population died during World War II, the Nazis never reached Tbilisi and its great pharmacy for the Red Army. The Soviets' remarkable use of improvised phage cocktails seem to have played a key part in helping a depleted and starving Red Army defeat a depleted, starving and disease-ravaged German army during the Battle of Stalingrad. The heroic efforts of Zinaida Yermolyeva, brewing anti-cholera phages in siege conditions in a bunker below the flattened city, helped keep Stalingrad and its defenders free from the cholera outbreaks sweeping through the German forces. The abysmal battle finally came to an end in January 1943, marking a turning point in the war, and Yermolyeva went on to become known as 'Madame Penicillin' for her efforts to help develop more conventional antibiotics in the Soviet Union. Together, Soviet phages and Allied antibiotics helped defeat the Nazis.

After the war, with the East and West ideologically and then physically separated by a line running through Berlin, an era of paranoid competition between communism and capitalism began. With perceptions of phage therapy already beginning to diverge, the regions' scientists were increasingly cut off from one another, and like newly separated bacterial cultures, they soon began to evolve in radically different ways.

In the new Soviet-controlled east, any kind of international collaboration with Western scientists became seen as *nizkopoklonstvo* – adulation of the West – and was made virtually illegal. When the Soviets' head of antibiotic production, Vil Zeifman, tried to secure the rights to Western-made penicillin instead of the failing *krustozin* in 1949, he was arrested, exiled to Siberia and later died of a heart attack under interrogation.[32] That same year, the Soviet Ministry of Health praised the work of the institute at Tbilisi, but researchers were chided for 'ideological lapses' such as acknowledging the past contributions of Western scientists to their field.[33] Preposterously, even the role of Felix d'Herelle was diminished, with a representative from the Ministry of Health declaring that 'everyone now knows that *our* scientist Gamaleya discovered the bacteriophage phenomenon'.[34] Nikolay Gamaleya was the celebrated Ukrainian doctor who had seen anthrax bacteria dissolved by a mysterious invisible agent in the 1890s, but who had later disputed the idea that phages were viruses.

As a 'golden age' of antibiotic development began in the West, leading to the development of at least forty different antibiotics, the Soviets continued to struggle to source or produce their own. Pharmacies might have just one or two types available, with some available one week and not the next. While the pharmaceutical companies mass-producing new antibiotics in the US generated vast profits to plough into further research and refinement of their manufacturing processes, the Soviet socialist model meant that its citizens' healthcare and the development of drugs simply cost the struggling state more and more money.

The insecure Stalinist regime could not accept its failure to provide such important medicines and began a propaganda drive to sow distrust in these scandalous Western chemicals. A campaign encouraged citizens to use home-grown, 'natural' Soviet products – including phages and all manner of dubious herbal remedies that were better for the Soviet people because they were *nashi*, meaning 'ours.'[35]

Unsurprisingly, the use of phage therapy in the Soviet Union therefore continued to expand in the decades after the war. In Soviet Poland, the Hirszfeld Institute of Immunology and Experimental Therapy began to look specifically at the effect of phage therapy on the immune system and began amassing a collection of hundreds of medically useful phages. More reliable commercial phage products in dried or tablet form were developed and, for years, millions of children in kindergartens and schools across the Soviet Union were given phages throughout 'diarrhoea season', a period from summer to autumn when cases of stomach bugs spiked. Phages were administered to sick newborn babies, with the treatment seen as less likely to upset their underdeveloped gut bacteria than the antibiotics that wiped out both good and bad bacteria. And, in one of the great achievements of Georgian phage science during this period, phage medicines were developed that were pure enough to be injected intravenously without causing fever-like immune reactions.[36]

By the 1960s, the Soviets were sending phages into space. Alongside two dogs, Belka and Strelka, forty mice, two rats, fruit flies, seeds, fungi, various bacterial strains and human cells, two types of bacteriophage (Escherichia coli T2 and E. coli aerogenes 1321) were blasted into orbit aboard the Korabl-Sputnik 2. An extended piece in *Pravda* summarised the experiments and described the phages as 'ultramicroscopic living creatures that parasitise bacteria and enter into complex genetic relationships with them'. Unlike the Soviet's canine space martyr Laika, who had been sent to her death in low orbit three years earlier, all the life on board returned to Earth, the bacteriophages presumably unharmed and still able to infect bacteria once back from their trip into the cosmos.

In the West, the development of antibiotics like penicillin had stemmed from a new, more rigorous approach to determining the effectiveness and safety of drugs. Medical groups began demanding that all drugs be developed and evaluated in a similar, standardised fashion in the hope of producing more compounds as reliably effective as penicillin, but for other diseases. The early iterations of what we now call the 'clinical trial' were born – a rigorous pathway of safety and efficacy evaluations on ever-larger groups of patients. Key in all this was the concept of a 'blinded' placebo-controlled trial, meaning the treatment was assessed alongside a control group who didn't receive the drug, and the patients didn't know who was receiving what until after the trial was complete. (In a 'double-blind' trial, now standard, the doctors don't know either, until the data is 'unblinded' and assessed later.) With no such studies on phage therapy – and instead a sea of conflicting reports and reviews – the treatment suddenly looked like a relic from medicine's dark and archaic past.*

It's not completely true that where antibiotics were available, they replaced phages: for a time, doctors in both the East and West prescribed both phages and antibiotics in combination, if both were available. Even Alexander Fleming's lab explored the use of phages and antibiotics together and found it to be a promising combination. (Today, this combined therapy is viewed as one of the most promising ways to fight bacterial infections in the future.) But as the convenience and

---

* In one of the rare large-scale tests of phage therapy from the Soviet era with a control group to compare the treatment to, 30,000 children across several Georgian neighbourhoods were split up into two groups, with the kids on one side of the street receiving phages and those on the opposite side receiving a placebo. The result was a 3.8-fold decrease in the incidence of dysentery with phage treatment. But it would be decades before these results were ever seen by Western eyes and, even then, they would be dismissed because other key information was not recorded.

effectiveness of penicillin and other similar drugs became more apparent, and its cost per course fell, enthusiasm for phage therapy fell away drastically in the West. By the end of the 1940s, the use of phages had sharply declined, with most attempts to market phage products discontinued. Published accounts of the use of phage therapy in Europe, the US and Brazil (which had at one point been among the most enthusiastic producers of phages in the world) had dwindled to virtually zero by 1949.[37]

As interest in phage therapy diminished, Western scientists instead increasingly saw phages as useful tools to understand the most fundamental molecular processes in biology. This new focus would lead to some of the biggest scientific breakthroughs of twentieth-century biology, paving the way for the new era of molecular biology and genetic engineering. But as decades passed, the idea of using phages in medicine was not just completely abandoned by the Western world; it was almost forgotten. Outside of the USSR, just a tiny number of practitioners continued to prescribe and produce phages on an 'artisanal scale' in France.

When the idea did occasionally resurface, it was remembered as unreliable, unscientific and economically preposterous. The Texan phage expert Ry Young, now a supporter of phage therapy, describes how his education in the 1970s instilled in him a belief that phage therapy was 'a bizarre chapter of medical history that should remain closed . . . and that their practical use in clinical application amounted to quackery'.[38] And besides, why would anybody want to use viruses to treat infectious disease when you have miracle drugs like penicillin?

For a time, antibiotics *were* true miracle drugs, revolutionising medicine and saving millions of lives a year. But life evolves, and chemicals do not. As none other than Sir Alexander Fleming wrote: 'The time may come when penicillin can be bought by anyone in the shops. Then there is the danger that the ignorant man may easily underdose himself and by exposing his microbes to non-lethal quantities of the drug make them resistant.'

As early as 1940, before it was sold in large quantities, studies of penicillin's effects on *Staphylococcus* had led to laboratory strains showing resistance to the drug.[39] Accepting the Nobel Prize for his discovery in 1945,[40] Fleming warned that 'it is not difficult to make microbes resistant to penicillin'. Hospitals and clinics, with vulnerable people coming and going in environments with high rates of infection and heavy antibiotic use, became hubs for the emergence of resistant strains.

The first strains of methicillin-resistant *Staphylococcus aureus* – methicillin having been invented to combat penicillin-resistant *Stapholococcus aureus* – began to appear in hospitals in the 1950s, just a decade or so after the 'miracle drug' was first made widely available. As each new 'miracle' antibiotic came onto the market throughout the twentieth century, strains of bacteria with high levels of resistance to them would appear soon after, on average only six years after coming to market.*

At first, the emergence of bacterial strains with resistance to a certain drug was interesting to bacteriologists, but wasn't

---

* Six years is the average time between the antibiotic coming to market and a resistance strain being *observed* by clinicians and their observations being published. In reality, resistant strains probably emerge far sooner, especially if the use of the antibiotic is widespread.

seen as a particularly pressing problem for healthcare professionals. If a person became infected with a microbe that had developed a resistance to one antibiotic, or resistance emerged during the course of his or her treatment, there was almost always another one available and new antibiotics with new modes of action – targeting entirely different bits of the bacterial metabolism or cell wall – were being developed regularly. Between 1940 and 1970, more than forty different antibiotics were developed and approved for use.

The discovery of these new classes of antibiotic gave false confidence that Western science and technology would always have an answer to antibiotic-resistant bacteria. As new pills flooded the market, antibiotic prescribing soared – with patients requesting them and doctors happy to offer them, often without even knowing if bacteria were the actual cause of the ailment. They also found use in agriculture and aquaculture too, where antibiotics were sprayed on fields or dumped into fish farms to prevent bacterial disease. When farmers realised that antibiotics not only helped control disease but boosted animals' growth, their use in animal husbandry exploded too. For a time, drug-resistant infections were a problem associated with hospital or care settings. But before too long, as the use of antibiotics became more wanton and extreme, it became an environmental problem: drug-resistant bacteria were developing and spreading through our waterways, our animals and our communities.

Anywhere between 30% and 90% of a course of antibiotics passes straight through the body and out through our poo, and so for decades, antibiotics have oozed into the environment down our toilets, our sinks and wastewater treatment works – or were splatted out of farmyard animals' backsides onto the soil to be washed into our rivers. The dosage of antibiotic compounds in the pills we consume may be measured in micrograms, but the amount of these drugs leaching into the environment each year is now measured by the tonne. A ticker from the statistics organisation *The World Counts* shows a live count of the estimated quantity of antibiotics used for livestock globally each year: just a third of the way through the year, the figure is 51,642 tonnes already, the figure changing every few seconds. The ticker is embedded in various websites about antimicrobial resistance, and so I keep seeing the figure rise ominously – by the summer of 2022 it's 60,000 tonnes. Attempts to curb antibiotic use through restrictions on prescribing, bans for non-critical use in agriculture, and education for doctors and patients about the dangers of resistance have largely failed to stop these and other ominous statistics ticking up and up to higher figures each year.

There are many ways bacteria can develop resistance to antibiotic drugs: they can evolve new metabolic pathways that chemically alter antibiotics, making them inert; evolve adaptations to prevent the uptake of the compound into the cell in the first place, or use tiny protein pumps to bail the toxic compounds out. When scientists talk of resistance 'emerging', it can sometimes sound like bacteria are constantly evolving clever new ways to counteract these antibiotics. Sometimes they do. But many types of bacteria already have strategies, evolved over millions of years, to help them resist the antibiotic compounds found in nature, like those secreted by *Penicillium* fungi. One of the oldest biomedical samples of bacteria ever taken and stored in the UK was recently found to be penicillin-resistant, yet it was taken from a soldier who had died of dysentery during World War I – over a decade before Fleming first identified the compound.[41] Bacteria with resistance to multiple antibiotics have also been found encased in a 13,000-year-old ice core.[42]

These adaptations normally come at a cost to bacteria, making them less efficient and therefore less competitive compared to other strains. But when we elevate levels of antibiotics in the environment, it makes these resistance mechanisms more valuable, and the genes for resistance become more dominant in the gene pool. Bacteria without the means to protect themselves from antibiotics are gradually replaced by those that can.

What's more, bacterial cells regularly trade genes with neighbouring bacteria, a process called horizontal gene transfer. This means that when a particularly useful resistance gene comes to the fore in a particular environment, it isn't just passed down generations of the same bacteria, it spreads *sideways*, throughout existing populations and sometimes between totally unrelated bacterial species.

Then there is the problem of biofilms. Many species of bacteria, as soon as they colonise an area of the body, begin to excrete thick, slimy substances which bind them and neighbouring cells together and to the surface they want to attach to. As the bacteria replicate and the colony grows, this becomes a complex, 3D-structure, which under a microscope looks like a kind of microbial city, with millions of cells packed together in undulating landscapes, crisscrossed by highways of connective fibres and pores linking individual cells. To give you an idea of how tough these microbial deposits can be, a multi-species biofilm we are all familiar with is tooth plaque. This sticky, acid-producing film can quickly form tartar, a calcified substance so hard it has to be chipped off by a dentist with a special pick. Antibiotics just can't penetrate through these tough physical barriers, especially if biofilms form in hard-to-reach areas like the lining of the lower lungs or the urinary tract. These stubborn infections mean long courses of antibiotics, which, in turn, further encourages the emergence of antibiotic-resistance within the biofilm.

Bacteria's aeons-old knack for survival, combined with our extravagant use of antibiotics, has combined to create today's horrendous populations of multidrug-resistant (MDR), extensively drug-resistant (XDR), and totally pan-drug-resistant (PDR) bacterial pathogens. Together they threaten to take us into a 'post-antibiotic era' – a world without antibiotics, where like the centuries of old, a simple insect bite or even just a graze can become a matter of life and death; and where many of our most important surgical procedures, from organ transplants to chemotherapy, suddenly become sure-fire routes to getting a deadly infection. This crisis threatens to dwarf the impact of any pandemic caused by a single bacterium or virus, and the Western world's scramble to find effective alternatives to the traditional chemical antibiotic are becoming increasingly desperate.

Of course, throughout the twentieth century, the Soviets had been working in a world where antibiotics either didn't exist or were in short supply. And so, when they encountered superbugs, they already had a solution to turn to. David Shrayer-Petrov, a Russian-American scientist and author, describes how in the 1970s, an outbreak of *Staphylococcus* with resistance to multiple antibiotics hit workers constructing a vast railway line between eastern Siberia and the Russian Far

East. Shrayer travelled with a specialist railway medical team sent to the region who immediately began using phages. In one particularly grim Russian town, Nizhneangarsk, residents lived in wooden huts with no plumbing or running water, with the nomadic railway workers living between them in crowded tents. Meltwater from the snowy, rat-infested streets flooded faeces and other waste from the residents' huts into the worker's tents, making the entire town a 'pathogenic cesspool', according to Shrayer. Up to 80% of the bacterial isolates sampled were resistant to multiple antibiotics.[43]

The team used cocktails of phages from Tbilisi 'applied with wads, sprinkles, washes and lotions' on the workers' pustulent abscesses and wounds, intravenous and intramuscular injections for more serious infections, and even the direct inhalation of phages for those with respiratory illnesses. Shrayer reported that their treatment was largely successful, with many people recovering with no side-effects or complications, even in those given phages intravenously every day for a week. But *yet again*: even if anyone in the West managed to hear about such a report, this was rough and ready mobile medicine, not a controlled clinical study.

By the late 1970s, the operation in Tbilisi had become vast. The institute employed a staggering 800 people to churn out thousands of litres of phage medicines in giant metal vats that operated twenty-four hours a day. A further 200 worked constantly to analyse the thousands of bacterial samples that continuously poured in from patients in clinics and hospitals monitored by the Soviet Ministry of Health. Additional phage manufacturing facilities could be found in the Soviet cities of Ufa, Alma Ata and Nizhny Novgorod.[44] The experts at Tbilisi travelled across the USSR, visiting far-flung cities and state-run factories so immense they had their own hospitals, taking samples and showing local medical staff how to administer phage therapies.

Scientists at the institute eventually worked out how to make phage cocktails pure enough to be injected directly into patients' bloodstreams without bacterial toxins causing severe allergic reactions and side-effects. The news of a new intravenous treatment for infections by *Pseudomonas aeruginosa* – a common bacteria that can colonise the body when vulnerable, for example in burns or open wounds, and can be stubbornly resistant to antibiotics – caught the attention of a senior official in the USSR's Ministry of Health, who was suffering from just such an infection. He declared he would be the first to be injected with the viruses. 'We could not refuse,' recalls Chanishvili, then a junior scientist. Her uncle, Teimuraz Chanishvili, was the director of the institute at the time, and she recalls how he went to the cinema after the minister had been treated. Halfway through the movie, government officials walked in and asked him to come with them. 'He thought he was being arrested for poisoning a government minister,' recalls Chanishvili. He had in fact been pulled out to be congratulated: his powerful patient was beginning to recover. 'He would tell the story with good humour, but knowing the story of Eliava, he was so frightened,' she says. Had the procedure not been a success, it could have been a very different story for Chanishvili's uncle and the repute of the institute.

The sweeping reforms enacted by Mikhail Gorbachev in the following decade encouraged the Soviet system to be less secretive and to acknowledge its recent bloody history, meaning that gradually, Georgians were able to discuss George

Eliava's life, death and legacy without fear. The building he founded was finally renamed in his honour in 1988, as the production of phages in Tbilisi peaked again. The Red Army was again at war, this time in Afghanistan; instances of anti-biotic resistance were on the increase, even in the USSR, where their use was more limited; and the production of a virus that can self-replicate into many trillion clones of itself in a few days was far cheaper than trying to synthesise sophisti-cated antibiotic chemicals in industrial quantities.

Meanwhile, in the West, an antibiotic called Daptomycin had recently come to market. It would effectively be the last: while variations of existing drugs have been created, no other new class of antibiotic has been produced since. Western scien-tists began their fruitless search for a new way to kill bacteria that continues to this day. The pharmaceutical industry, spooked by the idea of spending billions to develop a new drug that could soon become useless, or might sit unused – reserved only for the worst-case scenarios – began to divert investment away from antibiotics towards other drugs and treatments.

As the menace of antibiotic resistance became more serious across the globe, few if anyone in the West knew anything about the massive operation at Tbilisi and their alternative antibiotics. In the US in particular, phages had become key tools in the burgeoning number of labs focused on molecular biology and genetic engineering, but the idea that they could be used in medicine was not taught on any syllabuses.

Not *everybody* in the West had forgotten about phage ther-apy – pockets of scientists in many countries occasionally re-explored or re-evaluated the idea, against the grain of med-ical opinion, throughout the twentieth century, and a handful of doctors in France were still using phage therapy up until the

1990s. But the so-called 'red taint' – phage therapy's association with the sinister Soviet regime – ensured that when scientists did suggest that this form of medicine might be useful they were dismissed as eccentrics, cranks and, as the Cold War heightened, accused of being 'pinko commies'.

The expertise and growing bank of medically useful phages built up in communist Poland was just as hidden from the world as the operation in Georgia. A global antimicrobial resistance crisis was brewing and the group of people who knew how best to fight it just happened to be working in one of the most secretive places in the world.

Then, in 1991, after years of internal conflict and unrest in its constituent republics, the Soviet Union collapsed. President Mikhail Gorbachev resigned, as the upper chamber of the Supreme Soviet of the USSR formally declared that the USSR no longer existed. As the empire and its Iron Curtain came crashing down, and relations between Western governments and the Soviet system thawed, the free exchange of knowledge and collaboration became possible once again. But in the immediate aftershocks of the break-up of the Soviet Union, the Georgian scientists at Tbilisi had more immediate things on their mind: survival.

# PART 3

*Phage Fever*

# 7

## Coming in from the cold

No electricity. No heat. Broken windows. Weeds growing tall outside. The Eliava Institute in the early 1990s could scarcely have been a less convincing poster child for the exciting possibilities of phage therapy. As the USSR collapsed in 1991, Georgia declared independence from the failing central Soviet government, alongside other former Soviet states, from Armenia to Uzbekistan. After years of entwined, centrally managed economies across the USSR, Russia suddenly withdrew its funding, security and business, pulling the rug from under Georgian society. In Tbilisi, old ethnic tensions flared up within the economic and political vacuum, and the streets became dark and dangerous at night. Urban gun battles pockmarked the walls and shattered the glass of the capital. The bright lights of this once-glamorous city were replaced with the sinister gloom of candle-lit windows and the acrid smells of people burning anything to heat food and stay warm.

At the Eliava Institute, the red carpet that had once run up the grand central staircase was long gone, and the plaster on the walls had started to disintegrate. The Red Army's annual orders of two to three tonnes of phage medicines a year had been cancelled overnight, and the most valuable equipment was sold off. Staff would walk across the city to get to work only to find there was no power in the labs and, at one point, they even tried digging a well in the gardens of the institute in the hope of creating a reliable source of fresh water.[1]

Despite the hardship, there was still local demand for the viruses: soldiers engaged in civil conflict in Georgia's breakaway Abkhazia region were sent phage sprays that could treat five common infections in shrapnel and bullet wounds,[2] and most Georgians that could afford it had a packet of phages in their home first-aid kits for food poisoning and other stomach bugs.[3] The scientists at the Eliava Institute survived selling these products and small batches of phages to the remaining doctors who prescribed them, brewing them up and pouring them into tiny glass vials sealed shut with an open flame.

By 1993, the institute had fallen into total disrepair. The copper fermentation tanks that once brewed up hundreds of litres of concentrated phages now stood corroding among the sort of debris that gathers in a room with blown-out windows: litter, leaves and leaflets. The scientists moved their labs into the warmest and safest room in the sprawling institute, working with several dangerous pathogens from one table, with the rest of George Eliava's cathedral to phage science left to the elements. For a time, resources were so scarce the scientists resorted to using old vodka bottles to store reagents and an old Nescafé tin on a hot plate to melt the agar jelly used to grow bacteria on.[4]

During Georgia's long hot summers, the scientists struggled to keep the institute's unique archive of important viruses at the cool 4°C required to keep them stable. Colleagues had to persuade the staff of a nearby petrol station – one of the few buildings with stable electricity supply – to run a wire across the road from their forecourt over to the room where the scientists worked. Thanks to this, and staff who took turns to store the viruses in their fridges at home, many important phages were saved. But up to half of the institute's collection – some gathered and studied since d'Herelle and Eliava's time – may have been lost. During Georgia's icy winters, the phages themselves were safe, but the scientists who worked with them got ill from the cold. This famous institute, a repository of eight decades' worth of phage expertise, and one of the only remaining places conducting phage therapy in the world, was at its lowest ebb in its troubled history.

'I am not bragging when I say that I played a major role in people in the West learning about phage therapy in Georgia,' Professor Elizabeth 'Betty' Kutter tells me down a muffled phone line. There is some debate in the phage community about how best to address this remarkable woman, now in her eighties. The Godmother of Phage? The First Lady of Phage? Or simply the Queen of Phage?

Kutter has studied phages for over sixty of her eighty-two years on Earth. Broadcasting to phage scientists from as far afield as Mexico, Iran, Benin, Nigeria and Kazakhstan during the pandemic, from a big leather chair in her home in Olympia, Washington, like all good grandmas her webcam catches mostly ceiling and the top of her head. Although supposedly retired, Kutter remains a key figure in phage science, a matriarch who supports and inspires the informal and tight-knit phage community, especially female scientists early in their careers.

Her research has been led by her fascination with how T4 phages completely hijack the inner machinery of an *E.coli* cell, turning it into a phage production factory before instructing it to pop itself. The house she built with her husband was apparently designed to look, from above, like the distinctive icosahedral shape of a phage's head. 'This crazy phage is one of the loves of my life,' she comments.

Thriving in the highly competitive treadmill of academic research with young children remains a difficult feat even now but was virtually unheard of for a woman in the 1970s. Kutter jokes that her oldest son attended his first seminar at just ten days old, and spent the first three months of his life in a bouncer between the doorways of her lab. She recalls funders and senior male scientists being 'irate' that a woman with young children might even *apply* for permanent research positions at their faculties, with one even threatening to stop funding her grants.[5]

The sexism at America's bigger academic institutions led her to take up a position at the small but progressive Evergreen State College in Olympia, Washington, which in 1971 was just a few years old. Kutter has remained there, transforming this little-known college into one of the world's leading centres of phage research. Set in a thousand acres of evergreen forest, with its own organic farm and beach, the college is a liberal utopia that regularly makes headlines for its experimental approach – most notably the idea of sending all white students home for a day in 2017 to highlight racial inequality. Kutter's annual Evergreen Phage Meeting has become the most important event on the phage calendar, drawing scientists from all around the world to come together in a beautiful setting. Old and new generations of the phage community bunk down in simple dorms for a week and gather in Betty's garden to feast on a traditional salmon bake, cooked by local Chehalis, the native people of west Washington.

It was a chance trip to the USSR in the 1990s that led to Kutter playing a key role in the 'rediscovery' of phage therapy in Georgia. When the once mighty USSR fell apart and unrest broke out in its former states, the US diverted millions of dollars to programmes aiming to reduce the threat of nuclear, chemical and biological weapons in the region. American scientists were sent to Russia, Georgia, Kazakhstan and other former Soviet states to assess their scientific capabilities, build relations between Western and Soviet scientists, and ensure that the many scientists who were now without jobs did not seek employment in other rogue states.

Kutter found herself travelling to Moscow as part of an exchange programme organised by the Soviet and American Academies of Science.[6] Her work focused on her beloved T4, with not a thought for its potential use in medicine but to understand how a cell's genetic system can be so completely hijacked and reprogrammed by the phage's DNA. During her four-month stay in the Russian capital her laboratory colleagues kept talking about Georgia: how it was the most beautiful country in the former Soviet Union, with ancient towns set in picture-perfect mountains and incredible food. So, she and a friend took the 2,000km trip south for a week of hiking in the Caucasian country – which despite its troubles, could still offer breath-taking scenery and a remarkable cuisine of breads, stews, dumplings and cheeses, combining flavours from Greece and the Mediterranean, Turkey and Iran.

After falling in love with the country and its people, she soon returned to the region, and it was on her second trip that one of her new Georgian friends told her about a grand old building in Tbilisi that had once brewed phages by the tonne – for use as medicine. Intrigued, she decided to go and visit.

Kutter may well have been the first Western person to see inside the Eliava Institute since the start of the Cold War.

What she found there was not good. Georgia was still a poverty-stricken and wild place, and the scientists from the institute were desperately short of basic equipment and funding. Despite this, Kutter was amazed that Georgian doctors regularly used her beloved viruses to treat bacterial infections – and had done so for generations. The more she heard, the more excited she became about connecting the expertise of the Georgians with the facilities and funding available in the US. Once back home, she sent money for heaters and a generator to Tbilisi and was soon organising exchange programmes for Georgian scientists to study in the US, channelling research funding from Evergreen and the US National Institutes of Health to the struggling scientists at the Eliava Institute. She even sent money for taxis so its students didn't have to spend two hours walking to and from the institute every day.

In 1996, the science magazine *Discover* ran an article by the Russian-American journalist Peter Radetsky entitled 'The Good Virus'.* This seminal piece of reporting was the first to reveal to the American public that in a small pocket of the former Soviet Union, doctors were using a long-forgotten treatment that involved injecting patients with live viruses dredged from dirty water. It detailed the long and fascinating history of phage use in the Soviet era and how an American scientist, Betty Kutter, wanted to bring this old and derided form of medicine back to the states.[7] Radetsky and his brilliant

---

* A title that remains irresistible when writing about the power and potential of phages.

scoop helped bring the fascinating story of a forgotten therapy to a whole new audience, including a range of desperate patients suffering from chronic and untreatable infections across the world.

Others saw the *Discover* magazine article and sensed a business opportunity. One of them was property investor and venture capitalist Casey Harlingten, who happened to pick up the magazine while on a flight from Toronto to Seattle that year. Harlingten got in touch with Kutter, Teimuraz Chanishvili and Richard Honour, a microbiologist and biotech investor. Harlingten and Honour funded an initial round of research to see if phages from Tbilisi could help stop a particularly worrying strain of drug-resistant bacteria called VRE (vancomycin-resistant enterococci) killing patients on the other side of the world in Baltimore. Chanishvili soon found a phage in the Eliava Institute's archives which killed samples of the VRE sent over to Georgia from the US – in a test tube at least. It was enough to generate huge excitement among the investors. Soon, in the former communist party headquarters in the mountains above Tbilisi, the Americans were helping organise the first major international conference on phage therapy since the 1930s.

Unsurprisingly, the Georgian scientists who ran affairs at the Eliava Institute were not used to doing international business deals. Having worked in an economy carefully controlled by communist regimes for their entire careers, they were naturally suspicious of the American venture capitalist and his offer to effectively buy their expertise and archive of medically important phages. Would the $75,000[8] a year offered by Harlingten be turned into billions of profits by an American company and make them look stupid?

As the Georgians began to argue among themselves about how best to handle the deal, Harlingten's business partner started to have second thoughts. Having spent time in the ramshackle capital, with its sinister tower blocks and crumbling hospitals, and after seeing the dire conditions in which phage cocktails were manufactured, Honour began to doubt that the Georgians' products could ever be saleable. Plus, there was the problem of patents and ownership – if these phages were pulled out in jars from the city's sewers or the nearby river, what exactly did the Georgians have to sell? Did they really have anything that others couldn't learn to do themselves, with better equipment and facilities? And then there was the problem of how to market viruses from the former Soviet Union to American pharmacists and physicians.

The deal eventually fell apart, and Harlingten and Honour did exactly what the Georgians feared – set up a new company, aiming to genetically engineer phages so they had a broader host range – and, of course, so their products could be patented (unlike freely available, naturally occurring products). Kutter later told journalist Anna Kuchment she was embarrassed to have brought the venture capitalists to Georgia.[9] Nina Chanishvili would later tell American reporters: 'We gave the Americans access to all this background research, and they simply walked away with it. They told us we were stupid at business. Well, that at least was true.'

A few years later, Kutter was at a phage science conference in Montreal, when she got talking to an unusual-looking man who stood out from the crowd of academics, doctors and grad students. He was on crutches, smoking, and wearing a Greek fisherman's cap.[10] Alfred Gertler was a jazz bassist who had only recently learned what phages were. After breaking his ankle, Gertler had developed a deep infection in the joint that repeated surgeries and courses of antibiotics just could not shift and, in desperation, had found an article about a strange treatment that was still being used in some parts of Eastern Europe and the former Soviet Union. He would go on to become the first Western patient of many to make the difficult trip to Georgia to be treated with phages.[11]

Gertler's injury had been gruesome, but not the kind of thing that one imagines is beyond the will of the brightest minds in US medicine. In his forties, he had taken a job as a musician on a luxury cruise liner to help bring in some much needed cash for his two young kids. While in Caldera, on the Pacific coast of Costa Rica, he had decided to use some of his rare downtime to go for a hike in the lush hills overlooking the port. On his way back, rushing to make a rehearsal, he lost the footpath and found himself scrambling down a rather steep slope above the road. Grabbing onto a root to steady himself, he began to fall, and the root came away from the sandy soil. He dropped about 15ft. The bones at the bottom of the leg cracked and crumpled and split through the skin of his ankle. He endured an agonising ride to the nearest city, where his wounds were rinsed out and bandaged, and further agony along the country's bumpy dirt roads to the capital, San Jose, where his shattered ankle was put in a cast. Ironically, he'd refused a pain-killing injection for fear that it might cause an infection. By the time he returned to Toronto, however, the bacteria that would change his life forever were already multiplying deep inside the rock solid cast.

When the pain and swelling in his foot became excruciating, Canadian doctors cut open his cast and discovered the now rampant *Staphylococcus* infection. The prognosis was unexpectedly grim: lose the foot or lose your life. He tried another set of doctors in another hospital who told him the same. He refused to have the operation, and insisted they try and wipe out the *Staphylococci* in his bones with hardcore antibiotics. For years, he endured repeated surgeries and courses of antibiotics, but the infection remained, eating away at his bones and breaking out of the surface through two large weeping sores. He even spent a year with antibiotics being pushed through his bloodstream with an electric pump. Nothing worked, and his ankle was so delicate that he could barely get out of bed, let alone play music and support his young family. That's when he started to hang around at microbiology conferences.

As well as meeting Kutter, Gertler was introduced to several scientists interested in trying to help him. One of them was Revaz Adamia, head of a lab at the Eliava Institute and in Montreal thanks to Betty's efforts to connect East and West. Adamia told Gertler to send him a sample of bacteria, and he'd go back to Georgia and find a phage to kill it. It was that simple.

Ironically, phages had recently been used on a patient in Gertler's native Canada. Richard Honour had started his own small company in Seattle, which had supplied phages to a hospital in Toronto to treat a woman dying from a serious infection of the heart. The process worked, leaving her free from the infection, although she later died of an underlying genetic condition that had caused her heart problems in the first place[12]. But, neither Honour nor the woman's doctors had told Canada's health authorities about the procedure, let alone sought their regulatory approval – all parties had agreed to try the experimental treatment in complete secrecy. When the press found out, Honour was almost arrested for providing medicines that were not licenced for use. Gertler, desperately seeking a professional who could administer the phages being brewed for him in Georgia, tried for six months to contact the doctors involved, but they refused to speak to him – getting involved in phage therapy had almost cost them their careers. That left him one option: a trip to the destitute clinics in Tbilisi, Georgia.

In January 2001, Gertler, accompanied by Kutter, arrived in Tbilisi. They were collected by local scientists from the airport and taken through the city's pitted streets between various clinics and the local hospital, which was in even worse condition than the Eliava Institute. The lift was operated by an old beggar who charged a few Georgian tetri to get the shuddering metal box to the floor you needed to get to,[13] and refugees from the conflicts in Georgia's northern breakaway regions had taken shelter in some of the hospital's empty rooms. Cables hung from holes in the ceiling. In 2000, *The New York Times* reported on the still-miserable conditions within the Eliava Institute, where the scientific director 'shares a cramped and unheated lab with a fellow researcher, Marina Tediashvili, and half a dozen young medical students ... They are all shivering.'[14]

Gertler recalled to the science writer Thomas Häusler that in one freezing exam cubicle, an assistant tried to take a sample from his ankle – the injury angered and agonising after the long journey – with the end of an old coathanger.[15] Swabs were transferred over to the teams at the Eliava Institute, and as Gertler began to wonder whether amputation would be preferable to this ordeal, another local specialist came back with good news: several phages from the Eliava's historic collection had completely killed off a sample of the hardy *Staph* bacteria taken from his ankle. Waiting in the crazy hospital over the weekend for his treatment, Gertler asked Kutter to get him several bottles of vodka. Not for drinking, but to wipe down the icky surfaces around him so he could relax and get some rest.[16]

The treatment was intense. Pictures show Gertler propped up on a narrow bed in a dark and cramped hospital room, surrounded by medical equipment, a boiler and a sink. It looks like the 1970s, not the early 2000s. Surgeons flushed phage solution all the way through Gertler's damaged ankle bone, so much so that it went in one wound and came out of the one on the other side. They placed slivers of biodegradable material impregnated with phages deep inside his foot, and used a cocktail of other drugs and chemicals to break down the rotted flesh and put more pressure on the bacteria. Within a few days – for the first time since his cast was cut off in Toronto all those years back – doctors could no longer find any trace of the *Staphylococcus* in samples of fluid from Gertler's ankle.[17] He stayed in Tbilisi for a further two weeks to help his ankle recover and to ensure every last trace of the infection was gone. Within a few years, he was off crutches, back on stage with his bass guitar and walking to concerts.

Despite their lack of modern equipment and funding, the Georgians continued to find quiet success with their use of phages – treating a huge range of bacterial infections each year – and with their new connections to US institutions, began to test modernised products such as biodegradable, phage-impregnated bandages.[18] Despite being unable to advertise overseas due to the unregulated nature of their therapies, more and more patients made their way to Georgia, with dozens becoming hundreds a year, and in 2005 a small clinic in the western suburbs of Tbilisi opened with the help of American funding specifically for overseas patients.

The Georgians had become especially adept at treating burns which had become infected. In one particularly interesting case, three Georgian lumberjacks had been searching for a place to camp for the night in the snowy forests of western Georgia. When they came across two metal canisters that were warm to the touch, they thought it was a good stroke of luck and took them into their tents to huddle around like big hot-water bottles. Unfortunately for them, the canisters were the highly radioactive cores of a thermal generator from a Soviet plane that were still emitting radiation after being dumped in the woods many years ago. Over the next two days, two of the men experienced terrible redness and blistering of their skin from acute radiation burns. By the time they had been rushed to the capital, Tbilisi doctors found their weeping, dirty wounds were deeply infected with an antibiotic-resistant strain of *Staphylococcus aureus*. Following unsuccessful antibiotic treatment, a team at the Eliava Institute developed phage-impregnated sheets to cover the wounds. The discharge from the awful wounds, which had not been impacted by antibiotic treatment, decreased to virtually nothing after two days of the treatment. By day seven, the *Staphylococcus* could no longer be detected.[19]

To the Western scientific community, however, the intriguing case studies and magazine articles were not proof that phage therapy works. For that, you need carefully designed experiments involving lots of patients and control groups, to help discount the possibility that the positive outcomes occurred by chance, or by some other means. In 2005, the medical journal *The Lancet* ran an article on Georgia quoting several Western phage scientists who said the stories about phage therapy were 'hyperbole' and that carefully planned studies were required before it was seen as a solution for anti-microbial resistance. 'There are too many evangelists and too little data,' they wrote.[20]

Dr Alexander Sulakvelidze is a bear-like Georgian scientist now living in Columbia, Maryland. Known as 'Sandro' to friends and colleagues, his close-cropped white-grey hair and beard frame his stocky head and neck, with large dark patches sprouting from under the corners of his mouth. He tells me how in 1990, at just twenty-six, he was made head of the laboratories of Georgia's Centre for Disease Control. He had an entire floor of the centre's headquarters in the capital, a large team of scientists and plans to start the country's first molecular biology programme. But when the country descended into chaos following the break-up of the USSR, the simplest scientific work became impossible. 'Tbilisi in the early 1990s was nightmarish,' he says to me, shaking his head.

At the Centre for Disease Control, occasionally there would be enough equipment to help test samples from local hospitals for common pathogens or emerging diseases like HIV. But soon even that kit dried up, and the country's top scientists and public health officials could do nothing but sit there chatting and chain-smoking. He felt like his career was burning away along with the cigarettes.

'Day in and day out, we would just sit there and talk,' he recalls. 'Sometimes it was about what exciting science we could be doing if we had a certain reagent. But some of the time it was just discussing movies, or other routine things. It was really very depressing.' His greatest innovation during this period was diverting the water pipes that ran to his shower through his washing machine, so that the centrifugal force of the spinning drum created enough pressure to thrust water up and out of a shower head; he was tired of bathing with a bucket. 'It was an ingenious discovery, but not normal,' he laughs. 'It tells you a little about how people managed.'

Betty Kutter was not the only person in America trying to rekindle interest in using phages to treat bacterial infections. Sulakvelidze left Georgia in 1993, having been awarded a grant from the US National Academy of Sciences for students from the former Soviet Union to study at the University of Maryland. Here, he would be studying bacterial infections with Professor Glenn Morris, head of infectious disease at Baltimore's Veteran Affairs Medical Center. Soon after arriving in Baltimore, Sulakvelidze recalls finding Morris in a 'muddy mood'. A patient whose treatment for cancer was going well had become infected with vancomycin-resistant *Enterococcus*, or VRE. More and more of Morris's elderly patients were coming down with infections that did not respond to antibiotics, but this one was particularly disheartening. Just as the patient was on track to beat the cancer, what should have been a routine infection had suddenly killed them.

'Glenn was kind of deep in thought, and I asked him, did the phages not work?'

Sulakvelidze had in fact never worked with phages, even when in Georgia, but they were so commonly available in his home country that he assumed they were used in the US too. 'Glenn looked at me with eyes that said, "what are you talking about?"' Morris listened as his new Georgian student explained that in his country, they used viruses to treat bacterial infections – even drug-resistant ones – and that it often worked.

Morris was surprisingly receptive to the idea, Sulakvelidze recalls. 'Remember,' he tells me, wagging a finger, 'he is the guy who has to tell the families that their loved one just died. And so he saw the potential immediately. It was more a case of him persuading me that we had to get this technology across to the States.'

The organisations that fund and regulate medical research in the US were not ready, however, to take a chance on this long forgotten and strange form of medicine from the former Soviet Union. 'All our grant proposals were just tossed away,' says Sulakvelidze. 'People thought we were out of our minds and that this was just too ridiculous to take seriously.'

After repeated knock-backs from funders, the pair decided they would have to take the slow road to acceptance of the idea. In 1998 they formed Intralytix, a company focused on food safety. The idea was that if they could prove that phages work for killing dangerous bacteria in canned food, it would help them prove it could be used to kill dangerous bacteria in the human body too. Their first product, LystShield, targeted

*Listeria monocytogenes*, a food-born pathogen that while rare, has an extremely high mortality rate, killing as many as 30% of the people it infects. Further products targeted *Shigella* bacteria and *Salmonella* in food – bacteria which no industrial processes can 100% guarantee has been removed from products.

To their surprise, in 2006 the FDA approved the virus-based products as food additives, meaning they could be added to meat and even ready-to-eat foods. But the idea of using viruses in the food industry proved to be just as controversial as using phages in medicine. The media picked up on the story of 'viruses being sprayed on food' and went to town with it: television networks and news wires scrambled to get comment from 'experts' who had no understanding or experience of using phages to kill bacteria. The FDA had to create a frequently asked questions page on their website just because of the number of people asking about it. And it was even mentioned on an episode of iconic noughties TV hit *The Sopranos*. 'You know they spray virus on meat so they don't have to clean the rat shit out of the slaughterhouses?' Tony Soprano's troubled teenage son Anthony Jr says, ruining a family dinner of steak pizzaiola. 'And the FDA approved it.'[21]

The initial fervour surrounding the rediscovered 'Stalinist alternative to antibiotics' – as *The New York Times* put it – faded quickly. Bits of funding coming from Evergreen and Baltimore continued to help bring Georgian students to the US and to keep the lights on in the Eliava Institute – just about. But the country's best and brightest left for careers in the US and Europe, leaving a stretched and inexperienced generation of scientists in Tbilisi. There was still bitterness among the Georgian researchers about the failure to secure proper investment and jealousy of those who had received grants.[22]

More foreign patients were making the arduous journey to be treated with phages, but the treatments were still being conducted with rudimentary equipment and quality control that would make the regulatory authorities in the US or Europe shudder. Ten years after the collapse of the Soviet Union and the rediscovery of phage therapy in the West, there was little progress in developing a modernised version of the Georgian treatments that met standards expected by drug regulators and could progress to clinical trials. The 'second age' of phage therapy was encountering all the same problems as the first.

Harlingten and Honour's project to make genetically engineered phages in Seattle had also fizzled out, and Sulakvelidze and Morris's company continued to develop phage products for food processing and agricultural settings. Other phage companies founded off the initial buzz, such as GangaGen in Bangalore, also soon pivoted away from human applications to phage products for killing bacteria on farms or in food. The idea of using phages in human healthcare just seemed to be incompatible with Western healthcare systems.

'It was just so different, so unorthodox,' says Sulakvelidze. 'Nobody knew how the infrastructure would work, how the billing would work. There were just so many unknowns it seemed bound to die.'

# 8

## Keeping the faith

Despite the efforts of researchers like Betty Kutter and Alexander Sulakvelidze, entrepreneurs like Richard Honour and Casey Harlingten, and patients like Albert Gertler, the idea of phage therapy in the US remained stubbornly beyond even the fringes of mainstream medical science as the 2000s became the 2010s.

Across the West, there was still deep suspicion about Soviet science and what virologists in the former Soviet bloc were really up to. Professor Martha Clokie, a phage expert from the UK's University of Leicester with a long history of collaboration with Georgian and Russian scientists, recalls being asked by the British government to report back on what research she had seen in Tbilisi, amid concerns about potential bioweapons. Clokie's 'intelligence' actually led to the government funding a crucial translation of decades' worth of Georgian phage research into English.[23] Yet, she recalls that well into the 2010s, merely adding the words 'phage therapy' to applications for funding was still 'like a death sentence for a proposal' due to the ongoing perception of it as wacky by research funders.

Betty Kutter continued to sponsor Georgian science, and by 2012 the non-profit foundation she set up to promote phage therapy had directed over $100,000 to people and projects in

Georgia. Sulakvelidze continued to plug away with efforts to normalise the use of phages as antibacterials in other industries. But to most scientists in the US, a lack of good data on such treatments meant to try to use phage there was still an offence to modern evidence-based medicine.

To even begin to think about conducting a modern clinical trial, regulators like the FDA needed to see evidence that the substance being tested is well-understood, non-toxic, pure and stable, with a consistent formula. The phages used in the former Soviet bloc were none of those things.

Commercially available phage products in Georgia and Russia were and still are manufactured using so-called 'natural associations' of phages isolated together from water or soil – in other words, mixes of phages that combine to provide a broad spectrum of action against the most common strains of a particular bacterial species or several species. These mixtures are updated as frequently as every six months, meaning that unlike conventional drugs, with an exact chemical formula, phage products with the same name but produced at different times may contain different combinations and/or proportions of constituent phages. Genomic analysis of these products has revealed that they can contain up to thirty or forty different phages.[24] They are also not generally made under the strict standards expected in the West, known as GMP or 'good manufacturing practice'.

Alongside these general issues, biologists in the West had been taught that phages could only ever be useful as research tools in molecular biology, and little else. Almost one hundred years of efforts to make phage therapy a mainstream weapon against bacteria seemed, once again, to be coming to nothing. The regulatory hurdles and challenges of working with live viruses seemed to be insurmountable for Western physicians. Except, that is, for a few determined scientists who kept the faith – including a remarkably named podiatrist.

'I graduated before there was dirt,' says Dr Randy Fish, introducing himself to a virtual phage conference with a weary smile, shaking his head at the passing of time. He has broad, thin shoulders, smart white hair and a pale, serious face with hollows under his cheekbones. He is telling a captivated audience of doctors and virologists how he has been using phages to save homeless people's limbs for years.[25]

Forty years ago, while in his third year of medical school in Philadelphia, Fish found himself helping to treat what he called the 'Race Street irregulars' – a group of people in various states of addiction, destitution or homelessness living near his inner-city clinic. Many of them were suffering with terrible ulcerated wounds on their feet, the result of a combination of diabetes, poor hygiene and a life of pounding pavement.

Diabetic foot ulcers are not just abysmally painful and hard to treat, but they can make people very sick very quickly. Diabetes restricts blood supply to the extremities, meaning that even if a toe is amputated, that amputation wound may not receive the blood supply needed to heal. As the wound festers, further amputation procedures are needed, progressing horrifically upwards towards and past the knee. Treating a small sore on a toe can swiftly become an operation to save a limb.

Finding there was little literature on how to successfully treat such patients, Fish found his niche as a physician, specialising in podiatry and in particular, the treatment of diabetic foot ulcers. In the mid-1970s, Fish had attended Evergreen State College and at one point, been taught by Betty Kutter. By chance, years later, he was at a conference when a delegate asked a speaker about the possibility of using phages for difficult infections. Nobody knew anything about it. But Fish found the man and asked him where he'd heard of the idea. 'My neighbour,' he replied. 'Her name is Betty Kutter.' The chance encounter encouraged Fish to reconnect with Kutter, and the pair began to discuss the idea of using phages on diabetic foot ulcers. The wounds are often infected with multiple types of bacteria, but the dominant pathogen is *Staphylococcus aureus*, or simply 'staph'. Kutter supplied Fish with some of the vials of phage available commercially in Tbilisi, which contained a well-known and broad-spectrum phage that kills staph, known as SP1. All Fish needed to do was wait for a suitable case: a person 'up against the wall', with no remaining options but amputation.

In 2012, a patient with a particularly nasty wound came into Fish's clinic. The man had been walking on what was effectively a hole in his toe that went down right down to the bone, and an acute infection had taken hold over a weekend that wasn't responding to antibiotics. 'You could reach in and flip a piece of bone around in there,' says Fish matter-of-factly. He took the piece of bone out, flushed it out with the phage solution, and added more phage as he filled the cavity. From being a 'no hoper', the wound was looking better in less than twenty-four hours, and within two to three weeks of further treatment had healed over unexpectedly well with no surgery required.

Fish has been using phages to save people's feet ever since, when he can, applying liquid phage to the surface of flat ulcers, squirting it on medical gauze to dress wounds or dribbling it into more open wounds and joints with chronic bone infections.[26] He holds a clinic once a week, and so he reapplies the phage once a week too. It is not how dosage regimens are normally worked out, but it seems to work. 'Boy, it really cleans 'em up,' he says, like he's restoring parts from an old engine.

I ask how he got round the infamous regulatory hurdles that had prevented the use of phage therapy outside of the Soviet Union for decades. 'Well, we just kind of ignored them,' he says. 'God love the FDA, but we just went ahead and did it.' Fish has presented details of his work at the annual Evergreen Phage Meeting, where representatives from the FDA were present, and believes that because the treatment was so localised to the foot, applied in such small amounts and to desperately ill people, that the notoriously burdensome organisation was OK with him doing it. But it has since become more difficult for him to ship phages over from Georgia, with the authorities wanting to know they are being used for research, not medicine.

He now hopes to formalise his treatment regime to the point it can be tested in an official FDA-approved trial. But the same issues apply – any such test needs to be based around a product of a quality and purity that the FDA are happy with – starting with a therapeutic made to GMP standards. Shipping a box of unknown viruses in a brown parcel halfway round the world from a clinic in Tbilisi is not good manufacturing practice.

'This is really extraordinary,' comments a doctor attending the virtual conference, amazed at Fish's results on what are essentially untreatable bone infections. And it is. Yet it remains unavailable to most people with a diabetic foot ulcer, unless you happened to be one of the Race Street irregulars.

Fish was, in fact, not the only one of Kutter's Evergreen alumni using phage therapy in an unconventional setting in the US. Satya Ambrose graduated from Evergreen College in the 1970s, and for the past thirty years has been working as a naturopathic doctor – in other words, someone who only uses 'natural' products and therapies. At her clinics in Happy Valley, a pretty hillside suburb of Portland in Oregon, state regulations give naturopathic practitioners the right to use any natural medicinal product that has been approved for use in another country – for example, traditional Chinese herbal medicines – bypassing the regular drug regulations that govern mainstream medicine. Surprisingly, the phages produced in Georgia qualify as exactly this – a naturally sourced medicinal product approved for use in Georgia. The unlikely combination of Ambrose being a naturopath, being taught by Betty Kutter, and working in Oregon, means the Sunnyside Collaborative Clinic is also one of the tiny number of places in the world where you can receive phage therapy.

Ambrose has treated over a thousand patients with phages for problems ranging from chronic respiratory problems and skin infections to urinary tract infections. But here, phage therapy is offered in a clinic also offering contentious alternative medicines such as acupuncture, massage therapy and chiropractic treatment. Those hoping to rehabilitate the image of phage medicine might not be too keen on the naturopath loophole and an association with other treatments known to be pseudoscientific or lacking proof that they really work.

With just a few practitioners of phage therapy in the US and Europe, operating somewhat under the radar, interest in phage therapy began to plateau again in the 2010s. And the man who knew possibly more than anyone else about how to make it work in the US was tinkering with cars in his basement.

Dr Carl Merril, now in his eighties, has been interested in the idea of using phages to treat infections since the 1960s. With eerie echoes of what happened to Felix d'Herelle, the remarkable things Merril discovered about phages repeatedly brought him into conflict with not just his bosses but some of the most senior public health officials in the country, and he eventually took early retirement. His interest in phage therapy had meant he'd been swimming against an unforgiving tide his whole career.

Speaking with a high, hoarse, East Coast twang, Merril tells me how he attended the renowned phage science courses taught at Cold Spring Harbor Laboratory on Long Island in 1965. The courses are now famous for creating a generation of brilliant scientists who used phages to explore the most fundamental questions about what life is, essentially founding the field of molecular biology. Merril says two questions occurred to him while studying at the beautiful harbourside labs. One was, 'Why aren't we using these viruses to treat infectious diseases?' The other, 'How do we know these viruses don't affect us, directly?' Given that antibiotics were working well at the time, and had, in fact, saved Merril's life after a grave bout of septicaemia, he initially considered the first question less interesting than the second. So, he started to investigate whether phages really did only infect bacteria and nothing else – a pretty left-field question given that he was working at the National Institute of Mental Health at the time.

Soon, he was able to show, for the first time, that bacterial genes carried by a phage could be transferred into human cells, and the products of those genes might even be expressed by human cells.[27] The work, all at once, suggested that phage DNA could affect human cells, that human cells could be made to express microbial genes, and it even hinted that phages could one day be used to correct defective genes in genetic disorders. It also had rather dramatic implications for the idea of injecting phages into the body during phage therapy. Many were dubious of Merril's results, claiming they must be the result of sloppy methods and cross contamination between his phage samples and his human cell samples. But the debate over these important findings was quickly overshadowed by something else Merril had noticed during his experiments, something that would firmly establish him as a troublemaker in American medicine.

While trying to ensure he had no contaminants that could foul up his results, Merril had discovered that foetal calf serum – the supposedly sterile nutrient solution that laboratories use to grow cells in – was, in fact, often full of phages. This special serum, derived from the blood of livestock foetuses in slaughterhouses, was also one of the major components used in the manufacture of many vaccines. Merril studied eleven different lots of commercially available sera, and from all of them was able to grow a variety of different plaques, suggesting a multitude of different phages were present in all of them.[28]

Worried that phage contamination might affect the huge vaccination programmes of the era, Merril asked the FDA for samples of some common vaccines to investigate. They refused. So, as a registered physician, he bought a vial of measles vaccines from a local drugstore, writing a prescription supposedly for his son, Greg. He took it back to his lab and using the plating techniques developed by d'Herelle, found that every millilitre of the stuff contained many thousands of phages. He gave a public lecture on his findings, spelling out all he knew about the contaminated sera and vaccines; the possibility that phages can transmit their genes to human cells and the fact that phages can carry the genes for the toxins that cause diseases such as cholera and diphtheria. While he never suggested vaccines were dangerous, he suggested the FDA really ought to know which bacterial viruses are in which vaccines, study their effects, and aim to produce vaccines with as little contamination as possible, as soon as possible.

That may have been that, but for a science journalist called Gina Bari Kolata, who had been at Merril's lecture. She broke the story days later in the widely read journal *Science*, with an article entitled: 'Phage in Live Virus Vaccines: Are They Harmful to People?'[29] That, Merril says, was the beginning of the end of his career. 'When I came into work the next day, there was a group of people there who had on suits – and people in my lab generally don't wear suits. And they all looked very angry,' he recalls. 'The sentence I heard as I walked in the doors was that I was fired. I was dumbfounded – I didn't know who they were.'[30]

Merril believes the men in suits included figures from the FDA and possibly figures representing the pharmaceutical industry – although the FDA and his employer at the time, the National Institute of Mental Health, have not confirmed this. Merril wasn't actually fired – as a commissioned officer in the US Public Health Service he would have to be court-martialled instead – and so his employers tried to make him take a managerial role or take a sabbatical while defunding his lab, a move he says was designed to stifle him doing any further research on the topic.

The FDA eventually corroborated Merril's findings – finding that large numbers of phages were present in at least four vaccines which had been administered to over 600 million people in the previous decade. The manufacturers of these important mixtures of course regularly inspected them for the presence of bacteria, fungi and other infectious agents, but phages – the habitually overlooked phages – had completely slipped through the net.

A group of scientists ordered to review the problem by the FDA, in 1973, accepted that contamination with anything that is not meant to be in the vaccines was 'undesirable', but concluded that the phages were not an immediate threat to human health. 'F.D.A. Finds Four Vaccines Contaminated With Probably Harmless Viruses' ran a not-very reassuring *New York Times* headline.[31] The review concluded that the contaminated vaccines should be allowed to remain on the market.

Given federal regulations in the US forbade *any* extraneous material in vaccines at all, efforts to maintain the status quo ended with President Richard Nixon signing an executive order, remarkably, to allow vaccines to be contaminated with bacteriophages.[32] For all the wrong reasons, both the FDA and the US President had just essentially declared it safe to inject phages directly into the human body.

Despite his important findings, Merril found himself struggling for funding and derided by the scientific establishment for his bothersome views on phages and human health. In an attempt to prove the results of his earlier work was not due to sloppy work and contamination, he spent most of the 1980s developing a technique for detecting tiny quantities of gene products – that is, proteins – using silver ions, inspired by the silver chemicals he used in one of his other hobbies: photographing belly dancers. The technique, known as silver staining,[33] became a standard and patented way of detecting and tracking tiny amounts of protein in molecular biology research, making his employers a lot of money. 'All of a sudden, the National Institute of Mental Health decided I wasn't so bad and I was given funding for research again,' says Merril.

By the 1990s, the problem of antibiotic resistance in medicine was becoming more apparent and urgent, and Merril decided to explore his early questions about using phages to treat infections. In experiments involving phages and mice, Merril had found that the phages are actually cleared from the bloodstream quite rapidly by the mammalian reticuloendothelial system, or RES, a network of special cells for removing foreign objects. He began to wonder if this natural phage removal system could be a factor in why phage therapy had been so hit and miss throughout history. 'I woke up one morning and I thought, you know, I know how to make phage therapy better,' he recalls.

Working on mice infected with *E.coli*, Merril and his postdoctoral student Biswajit Biswas used Darwin's theory of natural selection to unnaturally select phages that seemed able to stay in the bloodstream for longer, injecting the phages into mice and then only propagating the ones that were still there after several hours. He repeatedly selected these long-lasting phages over and over until he had bred two variants of super-phage that could escape the RES for days. He named them Argo1 and Argo2, after Jason and the Argonauts. Around forty-eight hours after injection, the number of Argo phages still in the bloodstream was tens of thousands of times higher than the original natural variants. The mice also had a better chance of survival and fewer symptoms than mice treated with the natural variants.[34]

Soon after Merril's work on the long-lasting phages was published, an outbreak of drug-resistant infections at a nearby hospital became a deadly epidemic, and with echoes of d'Herelle, Merril saw it as an opportunity to put his theory into practice. Alongside long-time collaborator Sankar Adhya, he wrote a proposal to use phages to treat the patients.

'The review committee said, basically, "this is stupid",' recalls Merril. 'They said, "of all people to make this suggestion, Carl Merril should know that the bacteria is just going to become resistant to the phage, and so there's no sense in putting any effort into this."'

Merril did not have a chance to defend himself and point out the obvious: yes, populations of pathogenic bacteria can become resistant to both antibiotics and to phages, but at least when you get resistance to a phage, there are hundreds, thousands, possibly millions of other ones out there you can try.

Full of anger at the dismissal of his idea, he wrote a paper setting out his vision for what needed to happen to make phage therapy a reality in the West, published in the journal *Nature Drug Discovery*, in 2003.[35] As well as the use of his long-life phages, he set out the way new knowledge and twenty-first century technology could help address all of the problems that had dogged phage therapy over its long history, including bacterial resistance to phages. He called for more research to understand exactly how phages behave inside the human body; theorised on ways to ensure phages beat bacterial anti-phage defences; suggested ways to speed up the identification of bacterial strains and appropriate phages. He set out potential ways to broaden or change the host range of phages to suit new strains, and developed the idea of libraries of phages, already pre-screened for dangerous genes and approved by regulators.

Just a few years later, in 2005, a management committee recommended his lab be shut down. Amid budget cuts, the National Institute of Mental Health decided, perhaps unsurprisingly, that if savings needed to be made then they should probably come from the guy who has been writing about bacterial viruses instead of mental health for forty years. Over the next few years, Merril and Biswas tried to launch a suite of phage-based companies, but the idea of antibiotics made from naturally occurring (and therefore un-patentable) viruses remained a hard sell to investors. Biswas was hired by the US Navy's scientific research department, who by now were seeing a growing number of US servicemen were returning from conflict with drug-resistant infections. But Merril had had enough, and called time on his fifty-year career. He became the kind of guy who 'drove an Alfa Romeo while wearing a floppy hat', according to a profile of him published by the website STAT News,[36] swapping his passion for phages for photography, models, cars and other retirement hobbies.

However, this was not quite the end of the story for Merril. He and Biswajit Biswas would be reunited again in the most dramatic circumstances, as part of a high-profile case that has forced the medical profession to recognise that phage therapy, whether they like it or not, is sometimes the only treatment option they have left.

# 9

## Phages to the rescue

In 2015, a professor of psychiatry from San Diego, Tom Patterson, fell ill while on holiday with his wife in Egypt. Soon after crawling into a tiny, claustrophobic tomb at the Red Pyramid in Dahshur – where a guard warned him of the supposedly evil vapours inside – the burly, six-foot-five Patterson began to look ill and drawn. 'Just the heat,' he told his wife, Steffanie Strathdee, as they continued the tour and boarded a cruise up the Nile to visit more temples at Luxor and Karnak.[37] A professor herself, Strathdee was then director of the Global Public Health Institute at the University of California in San Diego (UCSD) and an international expert on a different kind of virus, HIV, and the social and behavioural risk factors that made it more likely that people would become infected with and spread the virus.

Between them the couple had visited some of the world's most remote and dangerous places, and the pair joked that Tom liked to collect weird infections and parasites like Pokémon cards. Despite their roles in public health, neither of them had much experience of antibiotic resistance. And unsurprisingly, neither of them had heard of phage therapy, since there was virtually no precedent for using it in modern American medical history.

After the visit to Dahshur, the couple went out onto the deck of their cruise ship for a romantic meal that they would eventually come to refer to as 'the last supper'. A few hours later, Patterson ran to the bathroom and vomited every morsel of his shellfish into the toilet of their small cabin. As seasoned travellers, they were used to picking up stomach bugs, and always packed a stash of 'cipros' – the all-purpose antibiotic ciprofloxacin. But Patterson continued to vomit throughout the night and next morning, and Strathdee was surprised when her husband said that he wanted to go straight home, without seeing the iconic tombs at their final destination, the Valley of the Kings.

Fortunately for Patterson, his wife had extensive connections to lean on for advice. She called Dr Robert 'Chip' Schooley, a professor of infectious diseases at UCSD and a personal friend who had helped them before when they returned to the States with drug-resistant boils on their arms after swimming through something rank in Goa, India. Despite Patterson's protestations, Schooley ordered them to call a doctor immediately, who was soon clambering into their cabin with a portable drip for fluids. He administered a stronger antibiotic, gentamicin. Patterson was still vomiting hours later and began to complain of back pain. With Schooley's help, Strathdee suspected possible pancreatitis. By the time the doctor returned, her husband was going into shock.

A lack of moorings at the dock meant their ship was tied to another ship, which itself was tied to another. It took eight people to transfer Patterson across the three ships to the dockside and to an awaiting ambulance. In the town's new and unfinished clinic, he woke in pain between doses of morphine and more antibiotics – this time cephalosporin, an antibiotic in the same family as penicillin. Nothing seemed able to halt the progress of Patterson's mystery illness, and twenty-four hours after their romantic dinner under the stars in Luxor, Strathdee could only watch in horror as nurses removed bags of murky green liquid that were draining out of her semi-conscious husband's stomach. While she made desperate calls to her health insurer to get a medical evacuation out of Egypt, Patterson became psychotic, first believing the doctors were experimenting on him, then screaming about an Egyptian colonel coming to kill him. Ironically, it was his wife that would end up turning him into a human guinea pig.

When a specialist team from Germany arrived, administered more antibiotics and stabilised Patterson for the six-hour flight to a military airstrip in Frankfurt, Strathdee noticed the paramedics repeatedly washing their hands and forearms carefully, which disturbed her. As soon as her husband was placed in his own room in the ICU of Frankfurt's Goethe University Hospital, a sign in German and English instructed nobody to enter without a protective gown and gloves. Was it to protect her sick husband from other infections and viruses in the hospital? Or to protect them from whatever had made him ill in Egypt?

Patterson was found to have not just acute pancreatitis – an infection of the organ that helps regulate the digestive system – but also a 'pseudocyst', essentially a lump of infected gunk which had formed around a blockage in his abdomen. Strathdee later described it as being the 'size of a football'. As if that wasn't bad enough, the bacteria causing the infection was none other than *Acinetobacter baumannii* – number one on the World Health Organization's list of the most concerning drug-resistant bacteria, or 'the worst bacteria on the planet', according to

Patterson's doctors. They had no idea when or how this awful bacterium had entered Patterson's body, but the strain was resistant to all but three antibiotics: meropenem, tigecycline and colistin, a World War II-era antibiotic known more or less as the official 'last resort' in the drugs cabinet due to its gruesome side-effects. The antibiotics given to Patterson so far had only succeeded in clearing out the infection's competitors and useful bacteria in his gut.

*A. baumannii* was once considered a relatively benign bacteria that generally coexisted with us and only threatens those with compromised immune systems. But it seems to be particularly adept at absorbing genes from other species – including genes that have helped other bacteria become resistant to antibiotics – and over decades, some strains of *A. baumannii* lurking in hospitals have become resistant to all the antibiotics modern medicine has in its arsenal. The bacteria is also called 'Iraqibacter', after thousands of soldiers serving in the Middle East in the 2000s picked up infections with it. Rumours spread that Iraqi insurgents were adding *A. baumannii*-laced excrement to explosive devices, helping to spread infections among the enemy. A more likely explanation is that the deadly strains of 'Iraqibacter' emerged due to overuse of antibiotics in overcrowded military hospitals, where they were then transferred back to medical centres in the US and Europe.

Strathdee, even with an entire career spent in infectious disease epidemiology, wrote later that the scope and scale of the antimicrobial resistance crisis had 'crept up on her'. Soon after Patterson was transferred to UCSD's Thornton Hospital in La Jolla, California, she was told that her husband's strain was now resistant to the last three antibiotics available. Pointing to the IV drips still delivering the antibiotics into her husband's bloodstream, she asked her friend Schooley 'what are those for then?'

'They're there to make us doctors feel better,' he replied.

Attempting to take her mind off the manic and relentlessly grim updates, she attended a virtual meeting with colleagues. When the meeting had finished, Strathdee heard a surgeon who had thought she was gone ask her colleagues quietly: 'Has anyone told Steff her husband is going to die?'

Patterson remained either semi-comatose or semi-delirious for weeks, then months. In the New Year, his immune system seemed to be rallying against the superbugs in his pseudocyst, but his delusions were worsening. He even called a family meeting to talk about euthanising him, having imagined a conversation with the doctor about his condition. Then, a drain that had been removing fluid from his cyst slipped, and the noxious infected fluid within flowed freely into his abdominal cavity. He went into septic shock, a dangerous immune response to bacteria in the blood, shivering so violently it shook the bed. He was put on a ventilator and into a medically induced coma. If things looked bad before, *A. baumannii* could now be found everywhere from his blood to his sputum. His whole body was being colonised by the world's worst bacteria.

With her husband now on a ventilator and his medical team fighting against ever more complications, Strathdee began to do her own research on what could be done to save Patterson from this seemingly undefeatable bug. Despite its fearsome reputation, Strathdee found a surprising lack of information on cases of *A. baumannii*, let alone detail on how to treat it. She found one research paper from 2013 on 'emerging therapy options for multi-drug resistant *A. baumannii*' and took a look at the list of new treatments in development: 'Ion chelation therapy', 'antimicrobial peptides', 'nitric-oxide therapy', 'photodynamic therapy', 'vaccination' and 'phage therapy'. The first two had only ever been tried on the bacteria in the lab and were way off use in humans. The next two could only be used on the skin. Vaccination was too late. That left the idea of using phages, viruses which Strathdee only vaguely remembered from her early years at college. She was puzzled to see that there was little modern literature on phages as antibacterials, aside from products approved by the FDA to kill bacteria in food, yet there were papers on phage therapy dating back to the 1930s and 40s, and reports of it regularly being used in Georgia, Russia and Poland. She read a *Buzzfeed* article about patients flying to a strange clinic in Tbilisi. Apart from that, nothing. Phages seemed to be the only realistic option to help her husband fight off his bacterial attackers, but she didn't know where to start.

The Declaration of Helsinki is a set of ethical principles, developed by the World Medical Association, to govern research and experiments that involve human subjects. It is an expansion of the Nuremberg Code, a similar set of rules developed after World War II, during which Nazi doctors conducted horrific experiments on people in concentration camps, supposedly to help advance wartime medicine. Following these horrors, and other unethical human studies conducted in the name of science, the Nuremberg Code and The Declaration of Helsinki were written to ensure that the wellbeing of individual research subjects always takes precedence over the interests

of science and society. The documents are both hugely important. But the final principle of The Declaration of Helsinki,
Clause 37, has proven especially relevant to phage therapy in
the last decade and a half. It states:

> In the treatment of an individual patient, where proven inter-
> ventions do not exist or other known interventions have been
> ineffective, the physician, after seeking expert advice, with
> informed consent from the patient or a legally authorised
> representative, may use an unproven intervention if in the
> physician's judgement it offers hope of saving life, re-establish-
> ing health or alleviating suffering.

In other words, if a patient has exhausted all known medical
treatment options and is experiencing great suffering or is very
likely to die, doctors may administer medical treatments that
have not yet been properly evaluated or proven to work. The
clause allows dying patients to try experimental drugs which
have not yet completed the long process of regulatory approval
and doctors to try treatments that are considered too radical or
risky for other patients. It's a long-winded, medico-legal way
of saying, when a patient is heading towards the abyss, 'it's
worth a shot'.

With attempts to prove the efficacy of phages still yet to get
off the ground, Clause 37 of the Helsinki Declaration has been
virtually the only way of getting phages into the arms of
patients outside of Georgia or Poland.

With her husband far too ill to even consider flying to
Tbilisi, Strathdee was desperately looking for someone with
knowledge of how to administer phage therapy. But aside from
the podiatry and naturopathy clinics, which at this point were

operating somewhat below the radar of the scientific establishment and medical press in the US, phage therapy had not been used in the USA since the 1940s. There had been some small-scale trials of phage therapy conducted outside of Georgia and Russia, but they were for rather different conditions, like Alexander Sulakvelidze's trial for the treatment of leg ulcers, and another in London on chronic ear infections.

It was enough to give Strathdee hope. After doing more digging, she found lab studies that showed phages existed that could kill *A. baumannii* in test tubes, Petri dishes, guinea pigs and rats. She found Dr Maia Merabishvili, a researcher from the Eliava Institute working in Brussels, who specialised in finding such phages. Strathdee's friend Schooley, now overseeing Patterson's case personally, was intrigued by her findings, and told her that if she could find phages with activity against *A. baumannii*, he would call the FDA to see if they would allow them to use it. Using the principles of the Declaration of Helsinki, they could apply for an exemption to normal FDA drug regulations known as an eIND – an Emergency Investigational New Drug licence – which would allow them to at least try.

'Our best hope was that the FDA would decide he was dying anyway,' wrote Strathdee in her book *The Perfect Predator*, which describes the battle to save her husband in agonising detail. The race was now on: not just to find a phage, or several phages, that could kill Patterson's strain of *A. baumannii*, but to fill in all the necessary paperwork to make the procedure legal. They would have to work out things like the dose and how to actually administer the phages later.

And so began a search of the world's phage research groups for a virus that could kill Patterson's strain of *A. baumannii*.

Unlike some other phages, which can work across a broad spectrum of strains and even across similar species, phages that infect *A. baumannii* are mostly 'type specific' – meaning they do not infect any old strain of the species – and therefore one would need to be found that was specific to the exact isolate infecting Patterson's body. Samples were sent to Ry Young, a professor at Texas A&M University, who Strathdee had seen quoted in a magazine as saying that finding a phage to infect any specific bacterium was 'relatively easy'. In many years of research Young's lab had found only a handful of *A. baumannii* phages though. He warned Strathdee that ideally, they'd want a handful of different phages, as the bacteria might become resistant to one type very quickly.

Meanwhile, back at UCSD, Schooley agreed to act as the principle investigator for this hastily organised experiment, and 'get the ducks lined up' – not just with the FDA but also the hospital's ethics committee. Even if a phage was found, would the legal approval come quickly enough?

As the number of researchers pulled into the case began to grow, Strathdee found there were more labs in the US working on phage therapy than she'd first thought – labs that didn't tend to publish what they're up to on public websites. The US Army and Navy both had dedicated units conducting research to advance phage therapy for the thousands of soldiers returning from the Middle East each year with terrible wounds compounded by drug-resistant infections. The Army was not keen on getting involved in a civilian case, but the Navy was willing to screen Patterson's sample for suitable phages in their collection, sourced from various festering bacterial hotspots around the world (including sewage and the wastewater or 'bilge' pumped from ships). Carl Merril's protégé, Biswajit

Biswas, was now chief of the Navy's Bacteriophage Science Division and had been continuing his mentor's life's work to understand how phages could be used in modern medicine. He had developed a high-tech piece of kit that could automate the process of working out if a phage was active against a strain of bacteria, meaning finding a match could take hours, rather than days. Plus, while Patterson was desperately ill, a team of doctors and researchers at Yale University had used phages, under an emergency compassionate use exemption, sourced from a New England lake called Dodge Pond, to save an elderly ophthalmologist from a catastrophic drug-resistant infection that had colonised his heart and chest cavity after a bypass operation.[38]

In February 2016, Patterson's kidneys started to fail and he required more invasive procedures just to keep him from starving to death. He had fought against the insertion of a new, larger feeding tube, even while in a coma, and was in wrist restraints. The procedure triggered a new wave of sepsis – complicated by a secondary fungal infection which had also breached the cyst and spread into his blood. They needed to find the phages fast.

By now, Strathdee had assembled quite a team. As well as Schooley, chief of infectious diseases at UCSD, there was Young, who had dedicated his career to understanding *lysis*; the way phages are able to burst bacteria from within like balloons. Approaching retirement, he was desperate to see some real-world good come from his years of research. There was Merabishvili in Belgium, who had experience of phage therapy from Georgia and worked with *A. baumannii* phages. And, of course, Biswas had worked on phage therapy since the 1990s with Merril, and both agreed to get involved. All of them had conducted phage therapy experiments on animals, but not humans.

Young and Merabishvili had just a few phages against *A. baumannii*, but Biswas and the US Navy had hundreds, sourced from some of the world's ickiest places by naval scientists. Teams at all the labs involved began working overtime to screen the phages they had for activity against Patterson's strain; they contacted other phage researchers who might have suitable phages, too. Meanwhile Strathdee and Schooley's paperwork, which might normally take a few weeks at best to be processed, was approved by the authorities – both the FDA and the hospital's ethics committee – within two days. It included documents making clear that those who provided the phages would not be held legally responsible if Patterson died.

In the middle of the night in March, an email from a member of Ry Young's team flickered into Strathdee's inbox with good news: they had found phages that could kill Patterson's strain of *A. baumannii*. Three new viruses had been picked up from the mucky floors of pig and cattle barns near Young's Texan lab, and one had come in from a phage start-up based in Strathdee's home city of San Diego. Then Biswas got in touch: he'd found ten phages in the US Navy's collection that worked on Patterson's strain and had selected the four most aggressive ones. After growing them in hefty conical flasks that held almost a gallon of bacterial broth, a long spin in a centrifuge helped concentrate the many trillions of phages from this raw mixture into a far smaller batch of cloudy phage-dense solution known as lysate. The phages were sent on ice by same-day delivery from Texas and the US Naval base at Fort Detrick, Maryland, to Strathdee, at her husband's bedside in San Diego.

But there was a catch – the same catch that had thwarted phage therapy in other cases in Europe already. The levels of bacterial toxins in the concentrated mixtures were too high for the FDA to let them be injected into Patterson – Helsinki Declaration or not. The phages had been propagated from trillions of burst bacterial cells, and therefore the solution was likely to contain fragments of those cells and other residues that could trigger a deadly immune response. The phages needed to be scrubbed clean first.

Rather than send both mixtures back to the labs they had come from, Strathdee and Schooley enlisted Forest Rohwer and Jeremy Barr at the nearby San Diego State University. Rohwer and Barr had become experts in purifying phages in order to study their effects on human cells precisely, and like so many others involved with the case, immediately cleared their benches of other projects and got to work.

While this agonising ad-hoc lab work was being organised, doctors told Strathdee that her husband was on the verge of multiple organ failure. As well as the compressors and ventilator keeping his heart and lungs working, and the coma protecting his brain, dialysis would soon be required to replace the function of his failing kidneys. While the phages were once again packed on ice and sent by courier to Rohwer's lab, several doctors who had barely even heard of phage therapy until a month ago now had to work out how to administer them: what dose, how often and where. There was barely any literature to help them get in the right ballpark. Dosage is a key aspect of medical treatment and is hard enough to get right with an inert chemical, let alone a living medicine that self-replicates more rapidly than any other form of life on Earth. Nobody knew the ideal dose to optimise their effect while minimising the risk of a dangerous immune response – a distinct possibility when injecting billions of foreign objects into the body, and when bacterial cells suddenly start being popped open and the toxic debris begins to course through the bloodstream.

Once again, a global consortium of experts made themselves available to provide advice, including phage therapy pioneer Merril. The team mixed expertise and instinct with details from wholly inadequate papers where phage therapy had worked in models using rats and mice. Then, there was good news from Forest and his team: in a blur of activity that might normally take them weeks, they had successfully purified the Texan phages. The process involved multiple rounds of filtration and centrifuging to try and chemically and physically separate the phages from the milieu of bacterial debris that surrounded them. The concentrated phages originally contained over 60,000 'endotoxin units' per millilitre, and the FDA required it to be under 1,000 to approve it for use. The latest reading was 667. Biswas and his team at the US Navy had performed a similarly impressive viral scrub, and his phages were also packed on ice and headed once more for the UCSD hospital.

With both sets of phages now pure enough for the FDA, the plan was to hit the world's worst bacteria from all angles: the 'Texan' phages from Ry Young would be injected repeatedly into the infected cavity in Patterson's abdomen, and if there was no immediate negative reaction the potent US Navy phages would be injected into his bloodstream. The double hit of multi-phage cocktails would reach all corners of Patterson's body and make it extremely unlikely that any cells could suddenly develop resistance to all eight of them at once.

After two months in intensive care, Patterson's body was failing and the race to get the phages into him was coming down to the wire. Strathdee finally had the paperwork from the FDA, giving her consent not just for the risky procedure but to allow experiments to be conducted on Patterson afterwards to better understand the procedure for future patients. When the phages reached the hospital, the purified mixtures had to be prepped by the hospital pharmacists – a tedious but necessary final step to ensure each was correctly labelled, correctly diluted and adjusted to the right pH for the place they were being injected into. Strathdee was told to go home, rather than watch the minutes tick by, and it was dusk when the call came through that the phages were ready.

A team of nephrologists sat grimly monitoring Patterson's kidney function, having delayed his much-needed dialysis in the hope the phages might help strengthen him for the procedure – but he was now rapidly deteriorating from kidney injury to total kidney failure. He could literally die at any moment.

As Strathdee made her way to the hospital via San Diego's snarled Highway 5, a team of ten doctors from across different specialties had assembled, all wanting to assist and witness this intriguingly left-field experiment, the medical equivalent of a 50-yard Hail Mary pass in the last second of a football game. Even Dr Cara Fiore, the FDA official who had processed the emergency approval, was keeping up to date with the situation by phone from her son's hockey match.[39]

Finally, the pharmacist entered Patterson's room with the all-important box, marked 'biohazard', and Strathdee's friend and 'holistic healer' blessed the phages as the doctors put them into syringes. Pictures were taken of the team with the phages to commemorate their exhausting and offbeat project, and then

Strathdee and Patterson's daughter whipped out her phone and blasted out a corny song to lighten the tense atmosphere: Survivor's 1980s belter 'The Moment of Truth'.

The 'moment of truth' proved oddly anticlimactic. The medical team didn't want to see signs of an immune overreaction, and it would be days of further doses before samples of fluid from Patterson's drains could provide data on how well the phages were working against the ever-mutating strain. But after three agonising days, there was some good news. Patterson roused from his coma, kissed his daughter's hand and drifted back off into an exhausted sleep. His recovery was rocky – as the phages spread through the bacteria and burst them open, his white blood cell count – a measure of his immune system's activity – initially spiked to dangerous levels as his body reacted to the vast amounts of bacterial debris coursing through his body. He suffered internal bleeds and organ failures that threatened his life once again. He even suffered sepsis caused by another bacteria, thankfully one that was still sensitive to antibiotics. And after rousing further and being taken off the ventilator, his *A. baumannii* infection roared back with resistance to the US Navy phages. Biswas and the Navy had to quickly find and purify another phage to finish the bacteria off. The new phage, small and circular and completely unlike the tailed phages used before, targeted a different part of the host and seemed to boost the activity of other phages and even some of the antibiotics. Eventually, slowly, Patterson made a full recovery.

As well as Patterson and Strathdee, Merril, Young, Biswas, Merabishvili and many other phage scientists around the world rejoiced. The drama became an international news story, recounted by Strathdee in her book and in many articles, TV reports and on social media, once again helping bring phages to the attention of the Western world and renew interest in this hundred-year-old medical treatment. Biswas, nicknamed 'the phage whisperer' among the group for his abilities, was soon giving a talk at the Pasteur Institute about the case. Among the audience that gave the talk a standing ovation was Dr Hubert Mazure, Felix d'Herelle's great grandson. 'That's when the calls started,' says Strathdee, who found herself acting as the de facto point of contact for critically ill people from all around the world hoping to access phage therapy.[40] The FDA began hosting workshops on the history and potential of phage therapies.

Patterson's dramatic recovery, and his medical team's incredible efforts, had provided a high-profile starting point for how to administer phage therapy in the US in a life or death context.

Sadly, not everyone who requests permission to use phage therapy through the emergency compassionate use route has been so lucky. Mallory Smith, an American writer and campaigner, died in 2017 at the age of twenty-five, two months after receiving a transplant to replace both of her lungs, which had been ravaged by cystic fibrosis. The genetic condition causes the build-up of sticky mucus in the lungs, digestive system and other organs, causing a wide range of symptoms and increasing the likelihood of stubborn, treatment-resistant infections. Diagnosed with the disorder at just three, by twelve Smith's lungs became infected with a type of bacteria known as *Burkholderia cepacia*. Somehow, it didn't stop her becoming a varsity athlete in high school and a club volleyball player at Stanford.[41] But during her sophomore year, the condition of her lungs deteriorated rapidly, and it became clear that she would need a lung transplant – a devastating procedure that has a high risk of complications and rejection. Sadly, after receiving a double lung transplant, the same bacteria simply recolonised her lungs, having hidden out in some other part of her chest cavity or upper airways untouched by the procedure.

Just as Strathdee had acted as the coordinator for an experimental treatment for her husband, it was Smith's father, Mark, who initiated the search for phages that might save his daughter's life. He reached out to Strathdee, the US Navy, and Carl Merril, who had by now set up a phage therapy company, Adaptive Phage Therapeutics. Hoping to repeat their remarkable feat, phage matches were found,* the necessary paperwork was completed, and the viruses were flown by a special medical plane and then helicopter to the University of Pittsburgh Medical Center, where Mallory was now critically ill. But it was too late: she suffered irreversible brain damage from her latest infection and died the day after receiving her first dose.

Mallory's mother, Diane Shader Smith, has become a 'phage therapy evangelist', giving hundreds of talks about her daughter's illness and how, if administered sooner, phages might have saved her life.[42] Her father Mark Smith has set up Mallory's Legacy Fund, which supports research projects and clinical trials that address antimicrobial resistance with a specific focus on phage therapy. Mallory's memoirs, *Salt in My Soul: An*

---

* Given the time constraints, the phages identified were not ideal for therapy or adequately purified, but Smith's parents asked that they be used anyway.

*Unfinished Life*, were published posthumously, and a documentary film about her life and death was released in 2022.

More and more compassionate-case uses of phage therapy have been conducted since Mallory's death, not just in the US, but also in the UK, France, Belgium, Canada, Australia, Japan and Italy. But these cases remain exceptional. More often than not, people with drug-resistant infections die without even hearing about phage therapy, let alone having the opportunity to try it.

The Helsinki Declaration, and the use of emergency regulatory approval for compassionate reasons, has proven to be a useful mechanism to save a hundred or so individuals from extremely perilous situations, but it is not the answer to a global crisis now affecting tens of millions of people every year. As one phage expert put it, regarding the process of gaining approval from drug regulators to use phage therapy for compassionate use: 'They never say no, but often the yes comes too late.'

# 10

## Desperation grows

Susanne de Goeij lives in a bright apartment in Amsterdam, full of colourful vases, trinkets and her young daughter's toys. With rosy cheeks and tousled, rich auburn hair, she looks a picture of health and happiness. But de Goeij's loose, casual clothes hide a painful problem. She has a skin condition known as hidradenitis suppurativa, or HS, caused by an overproduction of keratin, the fibrous protein that forms the structural base of our skin, hair and nails. The excess keratin gathers and pools in her hair follicles, forming what de Goeij describes as 'cement'. The hardened follicles swell up – often, particularly cruelly, on certain sensitive areas of the body, such as the groin, armpits and under the breasts and buttocks. Eventually, these follicles rupture, tearing open a wound from the deep skin to the surface. HS sufferers tend to have an unusual bacterial microbiome, and the wounds become infected and inflamed in a way that resists treatment. Each flare-up can feel like being sliced with a knife, and the pits and craters across De Goeij's back map the painful history of every single one.

De Goeij posts regular video blogs about her HS, sometimes in front of the planet wallpaper in her young daughter's room, sometimes while walking through her leafy local neighbourhood to her latest appointment. Her Dutch accent adds playful notes to English sentences, and de Goeij sounds happy, whether her update is good news or bad. Like most HS sufferers, she has tried many, many courses of different antibiotics, which tend to have anti-inflammatory effects for just a short while, before the effect fades and eventually stops. She has tried steroids, but they cause side-effects that mean they can't be taken long-term. Other treatments, which suppress inflammation by suppressing her entire immune system, she feels are too dangerous. 'I have a family. I'm not going to put my life on the line for this,' she says in her high, jovial voice.

The last resort for severe HS is surgery, to simply cut the areas containing swollen hair follicles out of the skin. But this is only really an option for those whose lumps are concentrated in a certain area, like around an armpit. De Goeij's cover half of her body. 'Every time they operate on some place, another one will just pop up on a totally different spot,' she says.

After years of failed treatments, de Goeij came across the high-profile story of Tom Patterson and Steffanie Strathdee, articles about phage therapy and the Eliava Institute's website. After researching the concept and making enquiries, she ended up crowd-funding the 8,000 euros needed for diagnosis, treatment, accommodation and the nine-hour flight to Tbilisi. With the Georgian capital then still regarded as a difficult place to reach and navigate, de Goeij used a 'patient advocate' in the Netherlands – a kind of phage therapy fixer who receives commission from the institute – to help organise the trip and liaise with the Georgian doctors. With abscesses on her buttocks, she endured agonising taxi trips across the city's gridlocked roads each day, and she now realises she could have found a cheaper hotel herself five minutes' walk from the institute.

De Goeij is just one of the ever-growing number of people now desperate for access to phages and phage therapy. The explosion of interest in it in recent years has arguably worsened the situation, with hundreds of articles just a few clicks away lauding the exciting potential of phages to save us from drug-resistant infections, yet just a few clinics are actually able to provide them – many of them only to people with the most dire prognoses. Although several clinical trials are underway for various infections in the US and Europe – from septic wounds to urinary tract infections – the patients eligible to be enrolled in these trials represent just a tiny fraction of patients out there suffering from chronic, untreatable, and sometimes terminal infections.

In Poland, which also has a long history of using phage therapy, the Phage Therapy Unit of the Hirszfeld Institute of Immunology and Experimental Therapy offers phage therapy for people who can get to its clinics in Wroclaw, Cracow or Czestochowa. The way they operate has become more limited since Poland joined the EU, and the criteria for determining which patients they accept is strict: they only treat patients with certain infections involving certain bacteria. With a lower profile and fewer international connections, even fewer people seem to know about phage therapy in Poland than phage therapy in Georgia. An expensive and complex trip to Tbilisi remains the only option for many wanting to try phage therapy.

What de Goeij found when she arrived in the city was, she says, 'behind in many ways'. Accustomed to using the latest silicon foam bandages, the clinicians used strips of gauze that you might find in an ancient first-aid kit. They took blood samples 'the old fashioned way' with a big syringe rather than a small finger prick. And she was to be treated in an old gynaecologist's chair, 'my feet in iron'.

'It was just a really weird experience,' she says, with admirable humour. 'I was lying completely naked on this very old gynae-cologist's chair and there were three Georgian doctors and a nurse standing looking directly at my vagina and talking, all without a word of English.' Facilities and language-barriers aside, de Goeij found the doctors' understanding of her condition to be excellent. As well as treating her HS, for good measure they gave her probiotic phages to help with a previously undiagnosed intestinal problem they believed she had. The trip and the treatment worked – her sores began to close up soon after her visit and her HS went into an unprecedented remission for three months. One of the institute's commercially available phage products had proven effective, so there was no need for a custom phage cocktail to be made specifically for her. Back in the Netherlands, she saved up for another pack of the phages to be sent to her so she could continue her treatment herself.

The Dutch postal service and government had other ideas, however. If Goeij attempts to get a parcel of Georgian phages shipped to her in Amsterdam, they will be seized and destroyed. Under EU rules, all medicines being shipped for use within Europe must be tested and approved by the European Medicines Agency (EMA). Even if the EU's stance on phage therapy changes, the EMA is unlikely to ever approve the complex mixtures of viruses made in Georgia – it is just too difficult to be sure what exactly is in them.

'They hold phages to the same standards as normal medicine, but it's not normal medicine,' argues de Goeij. 'As soon as the bacteria changes, the phage changes with it – this is its power.'

There are countless others who have endured long waits, difficult trips and bureaucratic challenges to access what they see as the only option for their chronic illnesses. Pranav Johri, after suffering the agony of a bacterial inflammation of his prostate gland for a year and a half, became the first ever patient from India treated at the Eliava Institute. He found it so difficult to get clear information on the subject and access to the therapy, that he and his wife, Apurva, have set up a company, Vitalis, in New Delhi,[43] specifically to help facilitate access to the treatment for patients in India. They have helped guide people through the Georgian phage therapy system – such is the demand for the service that they are now helping clients not just from India but the UK and Australia too.

A phage therapy clinic specifically for foreign patients has been operating in Tbilisi since 2005. The clinic's director, Dr Zemphira Alavidze, a charismatic semi-retired octogenarian who worked as a senior scientist in the Eliava Institute's phage biology department for fifty years, tells me of the many minor miracles they have performed on hundreds of people arriving in terrible states following years of failed courses of antibiotics, even during the COVID-19 pandemic. With the clinic's twenty-year-old website, and its inconspicuous brown metal door in a suburb in the far west of Tbilisi, it is not the kind of place that inspires confidence in your healthcare.

A new phage therapy treatment centre was added to the Eliava Institute in 2012, which can handle another few hundred foreign patients a year, and the institute now offers a remote service, where bacterial samples can be sent to their labs and suitable phages sent back. But as the demand for access to phage therapy grows, people are increasingly resorting to buying phages from pharmaceutical or even veterinary websites, and some are even trying to isolate and purify phages themselves. On a Facebook group called 'the Phage Page', scientists post links on the latest news about phages, but patients also share tips on how to get hold of phage products, including buying them from online marketplaces like eBay, or Europhages.com, a strange comic sans-strewn website that instructs people how to buy phages direct from foreign pharmacies. ('To find a Russian Pharmacy selling bacteriophage, click here, then copy and paste the word 'бактериофаг' (bacteriofag) in the search field,' it says.)

It's desperate stuff. The websites list places to buy phages in Russia, Georgia, Ukraine, the Czech Republic and Austria, with no detail on what viruses are actually in the mixtures. Beneath the article are dozens of heartbreaking comments:

'Hello. I am looking for matches for the following: Periodontitis and Lyme disease' . . .

'Can this eliminate enterococcus from my prostate?'

A man posting under the name *Mr Chestnut, Sales Manager* invites people to visit a website selling phages approved for animal use only.

Members of staff from the Eliava Institute post messages on the group page urging people not to try and source phages through these routes – '*the phages are not prescribed properly, the sensitivity test is not done and a doctor is not involved,*' one comments. '*We are also unaware of how the phages they provide are kept and transported.*' Even Betty Kutter, the grandmother of all things phage, chimes in to try and direct people to the few reputable routes to treatment available. But for many people, the prices charged by the Eliava Institute, even for its remote service, are just too high, and the service just too slow.

Even if things improve, scientists at the Eliava Institute or the Hirszfeld Institute in Poland, or a few naturopaths in the US – cannot treat every person in the world with a drug-resistant infection. The number of people waiting for access to such treatments grows ever larger.

In writing this book, almost every phage researcher I've spoken to – even those only distantly related to medical research – told me that they have received harrowing emails from the desperately ill and dying – emails that they can do little about. Some patients have been receiving a carousel of different antibiotic treatments for decades, only slowing the infection but not clearing it, with terrible knock-on effects as the bacteria they need to stay healthy are wiped out instead. One Australian researcher, Dr Jeremy Barr, has launched a project to study the testimony of antimicrobial resistance survivors, after receiving a message from a patient who planned to take their own life rather than go on suffering the daily pain and discomfort of a chronic infection.

With phage therapy still far from the mainstream, many of the world's pioneers in this field are professors and lecturers and doctors trying to push it forward on top of other research or caseloads, helping with phage therapy inquiries where they can in a voluntary or unpaid role, and occasionally dropping their plans to look for phages in their collections that might help a patient on the other side of the world.

De Goeij is eagerly awaiting her latest shipment from Georgia, albeit one that she must collect via Belgium, whose government has made a special exemption for therapeutic phages from the EU regulations. In the US, the FDA has also recently banned the import of Georgian and Russian phages.

'I think it is stupid of EMA not to step in and rethink this now, especially with antibiotic resistance on the rise,' she says, sounding less jovial. Three months' worth of phages – which she takes orally and rubs onto sores if they emerge – costs her over 2,000 euros. The Dutch health insurance system does not recognise packages of viruses made in Georgia as legitimate medical treatment and will only fund such treatment when there has been long-term, double-blind clinical trials for this particular disorder.

'That's going to take many years,' says de Goeij with an exasperated laugh.

Even more worryingly for people like de Goeij, the few research groups who have managed to test phage therapy in double-blind clinical trials have struggled to prove decisively that it is effective. There are many reasons why phage therapy does not fit easily into the traditional clinical trial, the first being that trials traditionally cannot begin until regulators see an exact and stable chemical formula for the drug being tested. Phage-based medicines, of course, often contain one or many different types of active and evolving virus, sometimes made bespoke for each and every patient, and they sometimes need to be changed during the course of the treatment if the bacteria develop resistance to them.

In a large-scale trial involving many patients, 'cocktails' of phages with many different types of phages included could be used to help ensure the same treatment works on a variety of patients and their slightly different bacterial strains; but the addition of multiple viruses only adds to the headache of getting regulatory approval. On the other hand, using a single phage strain is easier for regulators to assess and approve for the trial, but the likelihood it will work on different strains of bacteria in different patients in the trial is low, and the chance the bacteria will become resistant to it becomes much higher.

There are many, many other issues to work through: even if a mix of phages is made that regulators are happy with, the concentration of viable phages in it, known as its 'titre', can degrade quickly before a trial is completed, and phages can fall apart if administered into parts of the body with a pH or temperatures they aren't used to. Then there is the unpredictable nature of the body's immune response: in some animal studies the body seems to produce antibodies to neutralise the phages as if they are a threat; in others, the body rapidly excretes the phages from the bloodstream. These effects are likely to be different with different phages in different people. And what about dose? With a self-replicating entity that may or may not be cleared rapidly by the body, knowing exactly what dose to provide so there is a good amount of phages circulating around the site of infection is something of a guessing game. Regulators who assess the design of clinical trials do not like guessing games.

What's more, if a phage therapy project does get to the point of being tested, then traditionally 'phase I' of a clinical trial involves giving the treatment to healthy individuals, to test the safety of it. Given that healthy recipients will not have a bacterial infection, the phages do not replicate in the body as they would do during a real case, and so the resulting safety data is of barely any use at all.

Then there is the thorny issue of money: clinical trials are expensive, and traditionally pharmaceutical companies take the burden of the expense in the hope of reaping profits from the drugs that are proven to be safe and effective. In phage therapy, what is the product, exactly? Phages found in nature cannot be patented, and therefore are of no interest to a commercial company, as many who have tried to attract invest- ment for phage medical products will attest. Genetically engin- eered phages could be patented, but to develop a genetically engineered phage that works across lots of different strains of bacteria and different patients would take great investment, as Harlingten and Honour discovered when they attempted to do it in Seattle. That kind of interest and investment isn't going to come until there is more convincing data. It's a vicious circle.

Reviews of decades' worth of work in the former Soviet Union, and a growing number of compassionate-use cases in the West, all strongly suggest that phage therapy can be safe and that it is often successful. Phages are increasingly being used to control bacteria in industrial situations, and phage therapy has been deployed to help save a variety of animals from deadly bacterial infections, including livestock,[44] bees,[45] leatherhead turtles,[46] pet cats, dogs and reptiles,[47] and even the silk worms that produce the extremely sought-after Muga silk.[48] But agonisingly, trying to replicate this highly bespoke form of treatment within the rigid confines of a large, controlled study involving lots of patients treated together has so far failed to capture any significantly positive effects.

Decades after phage therapy was supposedly 'rediscovered' by Western medicine in the late 1990s, still just a few small research groups and pharmaceutical companies have been able to test the concept in modern clinical studies. Most have failed or been abandoned. In 2009, a pilot study of phage therapy for persistent ear infections showed promising results,[49] but the manufacturers decided not to invest in further trials, fearing that their product was unpatentable – again, because the phages were essentially unmodified from their natural state. The same year, Alexander Sulakvelidze's company Intralytix tested a mixed phage product for leg ulcers, targeting three pathogens at once in an attempt to make a product with a broad enough range of action to be of interest to investors – but it was found to be no better than placebo.[50] Several studies of phage cocktails to treat diarrhoea in children in Bangladesh in the 2010s saw no beneficial effect; probably because the phages being tested were not relevant to the local strains of bacteria in that particular location at that time.[51] And a scientist from a UK start-up that in 2010 made a pharmaceutical-grade phage product for 'nasal decolonisation' – that is, completely ridding the nose of bugs like *Staphylococcus* before surgery to help prevent infection of surgical wounds – told me 'we realised at some point that we had enough money to make another batch of phages, or do a clinical trial. But not both.'

In 2013, the European Commission announced it would be funding one of the largest trials of phage therapy in Europe since the 1930s, but it soon ran into familiar problems. With almost four million euros in funding, the 'Phagoburn' trial[52] was set to test phage cocktails for the treatment of infected burns, and was to involve several research groups, the French Ministry of Defence, and hundreds of patients from eleven military and civilian hospitals across France, Switzerland and Belgium. A French pharmaceutical company, Pherecydes Pharma, would find and develop two cocktails of phages isolated from the extremely fruity sewers running below Parisian hospitals, one targeting *E.coli* and one *Pseudomonas aeruginosa*.

However, the study hit two major snags soon after launching. First, doctors could not find nearly enough patients with the exact infections they wanted to test – most patients' burns were infected with more than one bacterial species, making them ineligible for the study. After looking for six months, only fifteen patients with pure *P. aeruginosa* infections were found; just one had a burn infected solely with *E.coli*. Second, the French medicines regulator, the National Agency for the Safety of Medicines and Health Products (ASNM), then ruled that Pherecydes Pharma must demonstrate that their cocktails of phages were of pharmaceutical-grade purity and stability: a standard requirement for assessing medicines containing one or two active chemical ingredients, but an extraordinarily difficult task when working with a mixture of living viruses that have been dredged from a sewer and liberated from concentrated bacterial cells. After delays caused by this and other regulatory challenges, there was not enough time to recruit a decent number of patients; the concentration of viable viruses in the phage cocktail was found to have significantly decreased, and the low dose helped bacteria develop resistance to the phages. Nightmare.

Arriving in Tbilisi now, there are few signs of the economic and political trouble of past decades. Although the Soviet-era tower blocks scattered on the slopes around the city look shabby and imposing, the capital is clean, modern and vibrant. White hybrid taxis hustle through six-lane highways and up and down the capital's steep winding streets, and on the wide, tree-lined avenues in the city centre, teenagers on e-scooters whizz past boutique shops, trendy cafes and casinos. The river Mtkvari, where George Eliava first found phages a hundred years ago, is now lined with gleaming municipal buildings and dramatic post-Soviet architecture, including an unfinished concert hall that looks like a giant piece of metal tubing has fallen from the sky.

As the Eliava Institute celebrates the one hundred and thirtieth anniversary of George Eliava's birth, the world seems once again to be forgetting the expertise and experience here in Tbilisi. Just one clinical trial has been conducted in Tbilisi, a collaboration with Swiss scientists from the University of Zurich and scientists from the Georgian National Urology Centre.[53] The trial failed to prove that the phage treatment was any better than standard antibiotic treatment or even placebo, although it did quite comprehensively prove that the administration of the Georgian phage mixtures was safe, at least. Nina Chanishvili tells me that the trial methodology was designed by her European collaborators and bore little resemblance to how doctors normally treat patients in Georgia. The Swiss team, having proven that phages are at least safe to use, are now working on new trials with their own phages and a large EU grant, but without the involvement of scientists in Georgia.

Despite the renewed interest in phage therapy around the world, and a steady stream of patients arriving to the Eliava Institute's clinic and pharmacy, the main research institute seems eerily quiet when I visit. Chanishvili, the head of research at this famed and storied place, says she currently has no research grant. Her latest research on phage therapy for treating asthma has come to an end, but she must wait for nine months to hear if the Eliava Foundation, which controls the confusing mix of research organisations and spin-off companies here, will provide a new round of funding for research.

Despite her expertise, she tells me that she is rarely contacted to help with cases or research pilots in other countries. A large section of the institute remains in the hands of a separate private phage production company, a legacy of when the manufacturing facilities here were privatised after the fall of the Soviet Union, and Chanishvili points to the mass of gleaming silver vents and pipes that have suddenly sprung out of the back of the building as the company attempts to manufacture phages according to Western GMP standards. 'They are the competition now,' she says, only half-joking.

Between 2012 and 2019, clinics in Tbilisi treated over 10,000 patients, over 1,500 of them foreign, from 71 different countries. A recent review found that generally speaking, less than a quarter of patients treated with customised phage treatments for infections with *Klebsiella oxytoca*, *Klebsiella pneumoniae*, *E.coli*, and *P. aeruginosa* suffered a recurrence of their infection.[54] Where infections recurred, the bacteria were often found to be resensitised to antibiotics, a result of the bacteria prioritising adaptations to prevent phages attacks at the cost of adaptations that help them fend off antibiotics. The institute continues to develop treatments for conditions never treated with phages before, such as Netherton syndrome, a rare hereditary disorder where babies are born with a strange membrane-like layer over their skin that causes recurrent skin infections.

But their successes still don't register with health organisations and investors wanting to see large, double-blind, placebo-controlled clinical trials, and they do not have the equipment to perform extensive DNA-sequencing of the bacteria and phages they work with, as is increasingly expected during microbiological studies elsewhere in the world. The Georgian phage mixtures and methods remain too variable and imprecise to be used or replicated overseas, so attempts to integrate their expertise into the strict Western clinical trial pipeline have proved expensive and unsuccessful. In Poland, another hotspot for phage expertise and research, a lack of funding has prevented any rigorous testing of their treatments too.

In recent years, Russia has also expanded its use of phage therapy, with millions of packets of phage products sold through the Russian pharmacological giant Microgen each year, and phages re-added to national clinical guidelines for antimicrobial therapy. It is even said that packets of phages are included in the healthcare regimen of Russian cosmonauts. But the country has once again become a global pariah thanks to Vladimir Putin's invasion of Ukraine, and calls to ban Russian scientists from publishing in major Western journals means any interesting developments there could once again be hidden from the rest of the world.

For now, patients are still coming to the Eliava Institute from countries all across the world, and the institute is still struggling to find the funds to modernise their facilities. The Georgians are a notoriously hospitable bunch, but I sense they are tired of telling the world what they can do. People are still 'discovering' the little-known Soviet treatment almost twenty-five years after journalists first reported from the Eliava Institute. Chanishvili has met countless journalists before me, and isn't hugely optimistic that the narrative will change any time soon. 'They are all still making the same article, the same documentary,' she says dryly.

# 11

## The third age of phage?

Rising up to the north of the French city of Lyon is the historic neighbourhood of Croix-Rousse, famed for its beautiful jumble of old labourers' dwellings and bistros painted in pastel shades of pink, orange and brown. Within this UNESCO world heritage site sits the rather less beautiful Hôpital de la Croix-Rousse, a sprawling medical campus made up of a mishmash of buildings, some of which look like grand old railway stations or hillside villas, and some of which look like multi-storey carparks. Here, behind the square concrete walls of Block H, I speak to patients with some of the most agonising and complex healthcare problems I have ever written about. But also, in Block H, a steady stream of patients are receiving phage therapy, performed as part of a carefully planned evidence-gathering study and supported by modern molecular and genetic technologies.

On top of terrible accidents and failed surgeries, these patients have suffered years and sometimes decades of recurrent infections, loss of mobility, open wounds, endless courses of antibiotics – and by the time they end up here, the spectre of amputation. Categorised as 'dead-end' cases by the doctors treating them, they come from all over France to see Dr Tristan Ferry, a specialist in bone and joint infections who will try to decide what, if anything, can be done to save their limbs. With thin-framed glasses perched over a blue mask, and grey hair spiked up slightly off his forehead, Ferry shows me pictures on his phone of some of his recent cases. Under the cruel light of an operating theatre, their battered flesh looks not unlike the various *saucissons* hanging in butchers' windows all over this ancient city.

Ferry's colleagues have begun to call his office the 'Musée de Ferry', or the Ferry Museum. On the bookshelves there is a beautiful old copy of *Arrowsmith*, the book inspired by the strange, conflicted career of Felix d'Herelle; a laminated print of an advert for d'Herelle's *Bacté-Instesti-phage* and *Bacté-Coli-phage* products made by his commercial lab in Paris; and several vials of Georgian and Russian phages sitting proudly alongside their original packaging. Ferry has become obsessed with understanding the viruses that he has started to use most weeks on patients in his clinic.

By his desk, in a polystyrene package, sits a giant bronze medallion of Professor Paul Sedallion, a doctor Ferry has discovered used phages on patients at this very hospital from the late 1950s until at least the 1970s, at a time when conventional narratives would suggest interest in this treatment in Europe was almost non-existent. Sedallion's entry in the *Journal de Médicine de Lyon* shows he was treating infections with resistance to antibiotics as long ago as 1959 – and was even 'training' his phages to be more aggressive against the bacteria. By the 1970s at least seventy patients a year with stubborn infections were being treated with phages, most likely found and cultivated from the wastewater of the hospital itself.

In Lyon, Ferry is leading efforts to establish modern protocols and systems to treat patients with phages not just in France, but, he hopes, across Europe. Good viruses can still be found in the sewers and beautiful rivers of Lyon, but things are a little different now to when Sedallion did his experiments here some fifty years ago. There are both the French and the European Union's stringent regulations to consider. To be accepted for use, even in 'dead-end' cases like those in Block H, the phages Ferry uses must be highly scrutinised first and be of super-high purity. This involves, at the very least, having their DNA sequenced and checked for toxic genes and going through the same extremely strict purification and quality control tests that have thwarted previous attempts to study phages.

In a shabby, virtually windowless basement on the other side of Lyon, a group of pharmacists collaborating with Ferry are starting to work out how they might create a library of well-studied and super-pure phages approved by regulators and ready for use.

'We have thirty years' experience working with chemicals,' says assistant pharmacist Camille Merienne, handing me the scrubs and shoe covers I'll need to enter the ultra-clean medicine production zone. 'But phages are completely different.'

Here, with strict decontamination procedures and pressurised rooms, specialist pharmacists prepare medicines for patients across the city's hospitals. Rooms used for pharmaceutical manufacture are normally high pressure, to ensure any contaminants flow out of the room not in; but in the case of phages, pharmacists must work in a low-pressure room to ensure the viruses don't escape – a 'biocontrol' measure required by law even though the phages are harmless to humans. In a lab back at the hospital, biomedical researchers show me Excel spreadsheets with information on the tens of thousands of genes in the phages they are working with: the vast majority are marked 'unknown' – genes which nobody has ever studied and so nobody has any idea what they might do. Regulators do not like unknowns.

Ferry has had to strike a delicate balance: encouraging France's only commercial phage producer, Pherecydes Pharma, to continue making the expensive, highly purified and screened pharmaceutical-grade phages that he can use for his patients, while also developing the same capacity in the hospital's non-profit pharmacy too. The agreement is, for now, that in every case he sees he will first check if Pherecydes' strains work first, before then calling on larger academic biobanks of phages, which his colleagues in the pharmacy can try to purify to the French regulator's exacting standards.

I ask him if he ever uses the Georgian and Russian phage products that sit on the shelves of his mini museum. 'No,' he says, grabbing one of the vials of liquid on his shelf. 'Phages are not meant to be yellow,' he says with a smile visible in his eyes above his mask.

Ferry, just like de Geoij, cannot use these products legally, despite them being readily available over the counter in Georgia and Russia, because it is impossible to tell the French authorities exactly what is in each batch. (The joke among phage scientists is that for the Russian phages in particular, even the products' manufacturer, the pharmaceutical giant Microgen, can't tell you what phages are in them. Or as one British expert tells me: 'they take all the viruses that infect a particular host and lob them in . . . it kind of works'.)

It currently takes Ferry several weeks to get the phages he needs, and some particularly difficult cases involve calls to phage researchers all across Europe and the Caucasus to find a match. Like Tom Patterson's remarkable case, which brought together experts from Russian biologists to the US Navy, treatments with phages too often still rely on a network of good-natured scientists putting in hours of work looking for phage matches at the bench for free.

As we speak, Ferry is eagerly awaiting a decision by the EMA on how it will classify phage-based medicines. Scientists in Belgium have managed to persuade their national regulators to treat phage products as 'magistral preparations' – essentially pharmaceutical concoctions made bespoke for each patient that are the subject of less regulation than marketable drugs. (Known as 'compounding' in the US, magistral preparations may be used to blend existing medicines, adjust the strength, make a pill into a cream, adjust the flavour, etc.) If the EMA creates a similar regulatory framework to Belgium, classifying each phage-based treatment as a magistral medicine rather than a new drug, it will make the administration of phages much easier across Europe. If not, it will mean anyone hoping to use and commercialise phage products must go through the same costly and time-consuming requirements for manufacturing and authorisation that are required for conventional medicinal products, making it virtually impossible for doctors to even test phage therapy unless they have access to pharmaceutical-grade phages. And there are unlikely to be many companies making pharmaceutical-quality phages unless doctors can conduct conclusive tests on the efficacy of phage therapy. The maddening loop continues.

Many see steering phage therapy into Belgium's unique 'magistral' legal framework as essential to the development of phage therapy in Europe, and possibly the rest of the world – and have formed a 'non-traditional antibacterials' group to pressure the EMA's decision. However, they may have strong opposition in the form of pharmaceutical companies, most of which would prefer regulators to classify phages as conventional drugs:

even for those companies not interested in developing phage drugs themselves, they sure as hell don't want people to be able to make innovative, cheap and unpatentable drugs without any of the regulatory hurdles that they have.

In a bright hospital room down the busy corridor of Block H, a tall former truck driver, known to me only as Mr B, sits in a T-shirt and underpants, dangling his thin, scarred and warped legs unselfconsciously over the side of a hospital bed. He is one of several patients to have received experimental treatment with phages who are in clinic today to check their progress. Through an interpreter, pointing at various parts of his body, Mr B talks me through the accident that changed his life almost forty years ago. In 1982 a truck jack-knifed in front of him and his truck, along with three other cars, ploughed straight into it. The bones in his legs 'exploded' from the impact, he says, miming a puff of dust with his hands. This was the start of an epic journey through France's healthcare system that typifies the miraculous medical procedures that we take for granted and which are all put at risk by drug-resistant bacteria.

Remarkably, Mr B's legs were painstakingly rebuilt with bone grafts and screws. But as the inner sanctum of his body was breached and probed and shuffled around by surgeons' hands many times over many years, he developed a terrible infection deep in his thigh. Bacteria spilled out into a disgusting open wound at the hip that according to Mr B, oozed and looked like 'meat'. Decades of strong antibiotics couldn't fully clear his *Staphylococcus* infection, but kept it dormant, until he banged his hip on a kitchen unit years later, in 2017. A further operation was needed and a deep and persistent infection reappeared within the hollow space running down the middle of bones in his right leg. This time, the infection was so bad he came to the attention of Dr Ferry and his clinic, for the most complex and persistent infections of bones and joints in France.

Ferry and his team opened up Mr B's infected leg, cleaned out the hollow part of the bones of his right femur with a whirring telescopic brush, and injected hundreds of trillions of phages directly into Mr B's bloodstream. (Ferry did not want to simply wash the bone out with phage solution, as such was the nature of the injuries he felt the phages would just pour out the other end.) Samples taken during the procedure revealed that twenty or so minutes after the injection, viruses could be found deep in the middle of the bones of his leg.

When I ask Mr B what he thought of the idea of having trillions of viruses injected into his bloodstream to treat his infection, he says with a shrug that although he has done some research, what phages are and what they do is really too complex for him to fully understand. 'It was like he was talking Chinese at first,' he explains through the interpreter. 'But I'd had so many experimental treatments, I thought, "Why not?"' Mr B is in the clinic today for a check-up, and it appears his phage treatment has worked. Although he still has pain in his knees and walks with a stick, the infection in his upper leg has cleared and the wound on his hip has healed over. He is now able to work as a live-in groundsman tending to the home of a wealthy local family.

Back in Ferry's office, the doctor shows me a heavy box packed with correspondence from people across France who have contacted him seeking treatment with phages – for all manner of problems, not just bone and joint infections. When I ask how he feels about so much unmet need, he is not despondent – each successful case he completes is a leg and a life saved and a step towards acceptance of this old medical concept – first pioneered here in France of course, over a hundred years ago.

Ferry's work is not just as a bone infection specialist: his role has become a project manager, salesman, lobbyist – helping to coordinate and navigate the various scientific, economic and regulatory challenges of every case. He hopes that one day Europe will have a network of dedicated phage therapy centres in every country, each assessing which local patient cases are most well-suited to phage therapy and working together to source, manufacture and share EMA-approved phages for clinical use. His energy and enthusiasm are infectious, and despite the endless challenges of working with viruses, I leave believing that there will be plenty more people like Mr B in the future, walking out of Block H down the beautiful slopes of Croix-Rousse – and back to the lives they had before they encountered drug-resistant bacteria.

Since I started writing this book, in 2020, the number of phage therapy cases, studies, and research projects being announced has suddenly started to surge again. In just the space of a few months there has been a flurry of reports that demonstrate successful phage therapy. At a Belgian military hospital phage therapy has been used to save a victim of the 2016 Brussels bombing attack, whose repeated surgeries and skin grafts had become hopelessly and chronically infected with *Klebsiella pneumoniae*;[55] then there's been news of a fifty-six-year-old immunocompromised patient being successfully saved from a flesh-eating *Mycobacterium chelonae* infection with phages at a Harvard teaching hospital.[56] There are publications on genetically engineered phages being used to clear a severe *Mycobacterium abscessus* infection caused by a patient's cystic fibrosis;[57] and news that scientists and doctors in

Australia have developed a framework to standardise how phage therapy is accessed and administered across the country.[58]

At the time of writing, there are almost a dozen modern double-blind clinical trials of phage therapy either being planned or underway, involving both cocktails and single-phage preparations, natural and engineered phages, for a suite of different bacterial problems – from *E.coli* urinary tract infections and *P. aeruginosa* in burns to multidrug-resistant *S. aureus* infections of the blood, diabetic foot ulcers and infections associated with cystic fibrosis. As well as the work on prosthetic joint and bone infections mentioned above, there are phage treatments at pre-clinical trial stage for the treatment of *Shigella flexneri* (dysentery), atopic dermatitis, acne, tooth decay and for the treatment of people with bacterial respiratory infections exacerbated by COVID-19.

More fundamental research is looking at the potential for phage therapy in illnesses such as leprosy, TB, chronic obstructive pulmonary disease (COPD) and diseases caused or associated with bacterial imbalances, such as hidradenitis suppurativa, inflammatory bowel disease, asthma, Crohn's disease, ulcerative colitis and even certain types of colon cancer associated with bacteria. Phages are being explored as a way to not just treat but also detect certain pathogens, for example in the rapid diagnosis of sepsis.

Steffanie Strathdee, after saving her husband's life with phages, has helped set up the Center for Innovative Phage Applications and Therapeutics, or IPATH, the first such centre of its kind in the US. The Canadian phage researcher Dr Jess Sacher, together with web developer and engineer Jan Zheng, has set up the Phage Directory, a network dedicated to connecting phage patients to phage researchers and practitioners, with an alert service to help doctors issue emergency phage requests to phage labs worldwide for patients with life-threatening infections. The directory has so far helped coordinate forty cases, including two successful treatments for young children and one for a leatherhead turtle.*

'Everything has changed in the past few years,' says Sacher. She remembers, as recently as 2017, being embarrassed when she told other scientists she was an advocate for phage therapy. 'I was always told by everyone around me that phage therapy was never going to work. It was always something we put in the intro paragraph of our grants but was not to be acted upon for real or you would look crazy.'

As of 2022, Sacher's directory now has hundreds of subscribers from hundreds of phage laboratories across the world, and can direct patients and doctors to at least four different centres that offer phage therapy in the US, Belgium and Poland, rather than just directing people to the Eliava Institute website. Sacher and Zheng are also working on what they call an 'operating system' for phage therapy, a kind of user-friendly platform for helping patients and doctors navigate the logistical, regulatory and medical challenges of phage therapy in a streamlined and consistent way.

In the long and turbulent history of attempts to make phage therapy work, there has never been so much good news, interest and momentum. The world's largest research-funding organisations, having once dismissed phage therapy proposals

---

* A Twitter post on New Year's Eve appealing for phages to help save a forty-two-year-old female leatherhead turtle with an infection in its shell known as septicemic cutaneous ulcerative disease, or SCUD, was one of the directory's most popular tweets ever. 'People love to help turtles!' Sacher told me.

just from the title alone, are now dishing out millions of dollars in funding each year to explore both phage therapies and the fundamentals of how phages work. There are now fifteen official national 'biobanks' collecting and studying medically useful phages in countries from Canada to Korea. In the UK, in 2022 the National Health Service appointed its first 'clinical phage specialist' to drive forward the use of phages in hospitals, and MPs from the House of Commons Science and Technology Committee launched an inquiry into the use of bacteriophages as an alternative to antibiotics.* At least fifty biotechnology, pharmaceutical or life science companies are now developing phage products, including phage-based antibiotics for live-stock, crops, and fish-farms, phage-based vaccines, phage-based probiotics, phage-based pest control products and phage-based products for disinfecting surfaces, for rapidly detecting dangerous bacteria, even for preventing the corrosion of oil and gas wells and pipelines. The list of new ideas, projects and organisations grows almost every week. The current glut of experimental case studies and clinical trials may hopefully soon produce much-needed data on how well different phage therapies work for decent-sized groups of patients, in a range of different infections and approaches.

The road to widespread regulatory approval and availability will likely stretch on for decades, however. There remains lots of data to gather, many lessons to learn, and there are still unanswered questions about how to organise and fund such

---

* The decision to focus an inquiry on bacteriophages came after members of the public were asked to submit ideas for the committee to investigate. An investigation into bacteriophages as an alternative to antibiotics was chosen as the winning entry from over 90 submissions.

complex treatments in the different healthcare settings and systems found in different parts of the world.

Even if the results of the latest clinical trials are positive, the same old questions remain: how will regulators approve a treatment when the actual mix of viruses used may be different for each patient? Do we really know enough about the complex, cell-popping, gene-swapping behaviour of phages to use them regularly in our bodies? Should cocktails of phages be developed that might work on a wide range of patients with similar conditions, or should each and every patient have a bespoke set of phages found for their particular strain of bacteria? Do doctors really understand what will happen when phages are injected into the body, and how should they respond when resistance to the phages they are using emerges? How can such complex and labour-intensive medicine ever be available to millions of people?

We will return to these questions, and to what the future of phage therapy may look like, at the end of this book. There is so much more that phages do for us, beyond medicine, that is to be appreciated: from helping regulate our digestive microbes to regulating the planet's oxygen levels and climate, the important part they have played in the evolution of complex life on Earth, and their starring role in some of the biggest biological breakthroughs of the past one hundred years. We'll look at all the phages out there in the world, and some of the amazing things we've learned from just the tiny fraction that we've studied so far. And there are even more bold, brilliant and eccentric figures from phage science to celebrate, without whom we might not know half as much as we do about how life on this planet works.

# PART 4

*Fundamental Phages*

# 12

## Atoms of biology

One day in 1937, a German physicist called Max Delbrück missed a lecture he'd been keen to attend. Emory Ellis, a professor at the California Institute of Technology (known as Caltech), and one of just a handful of people in the United States studying bacteriophages after almost two decades of squabbling over what they were and who discovered them first, was giving an introduction to his work on the viruses. Delbrück, having only just settled in the States to take up a post at Caltech, was annoyed at having missed the lecture, and so skipped across the Pasadena campus to find the professor in his lab.

Ellis was researching phages in the hope that they might help shed light on the relationship between viruses and cancer. A biochemist rather than a microbiologist, he had effectively taught himself the basic techniques d'Herelle had employed, using only rudimentary microbiology equipment – little more than a few dozen glass plates, a few dozen pipettes, and an autoclave (a glorified oven for sterilising equipment). But with this basic equipment, a litre of wastewater from the Los Angeles Sewage Department, and some fiendish mathematics, Ellis could calculate numerically precise details of the relationship between a virus and the host that helped it to replicate, just like d'Herelle had.

Arriving in Ellis's lab, Delbrück saw the same glassy holes scorched into a plate of bacteria that had mesmerised d'Herelle and Twort at the beginning of the century. And when Ellis explained to him how those plaques on the plate could be used to derive information about *individual* viruses, including how long it takes them to replicate and how many burst out of each bacterial host, Delbrück was astonished. 'I don't believe it!' he cried, which would become his go-to catchphrase whenever he heard some bold new scientific claim.[1] After checking Ellis's numbers, he did believe it alright. And he believed he was staring at the key to unlocking some of the biggest questions in science.

Delbrück was one of a number of talented physicists that decade who had switched fields and become interested in biology, which had become somewhat bogged down during the inter-war period, as biologists struggled to make progress on the grand questions of life. Specifically, how exactly living organisms pass on a blueprint for life to their progeny, and how the progeny then builds itself from that blueprint. Of course, we know now that this is possible thanks to that famously clever staircase-shaped chemical called deoxyribonucleic acid, or DNA, which encodes information in the form of a string of chemical letters, which are translated by cells into the thousands of different proteins required for cells and tissues and organs and bodies to function.

However, as of the 1930s, DNA was just one of many chemicals that had been found within living organisms whose exact structure and function was a complete mystery. Biologists had made progress understanding the rules of inheritance and reproduction – for example, the ratios and frequencies that certain characteristics were passed on through the generations

– and they called the various competing units of inheritance *genes*. But with no understanding of DNA and the other complex molecules that help it to function, they had no idea how exactly the individual organisms transmit the information about their nature to future generations. What were genes *made of*?

Biologists had become expert at crossbreeding simple plants or creating broken, mutant variants of flies, bombarding these organisms with radiation and observing the myriad ways the offspring came out with their genes mixed up and their bodies messed up. They painstakingly catalogued the various mutant strains and the genes associated with particular characteristics. But they weren't getting any closer to understanding the hidden process that they were disrupting. Biochemists broke open cells and worked through the thousands of colourless compounds inside, in the hope of understanding the special chemical reaction at the heart of it all. The more they looked, the more complexity they found. On the great central question of how life builds itself, and how it propels the beautiful intricacy of itself from one generation to the next, biologists were flummoxed.

When great thinkers did occasionally come close to imagining how a biological chemical might be able to carry information and replicate itself, their ideas fizzled out. They were far-fetched, speculative theories with no way of ever being tested. Many of those who did consider such questions presumed protein was the genetic material,[2] not DNA – it made sense given that these chemicals could come in endless different forms, much like the seemingly limitless variety of characteristics seen among life on Earth. That's when the great theoretical physicists of the era began to believe there was something more fundamental at the heart of it all – and that perhaps, an entire set of new physical laws was required to describe the living world. But the typical methods and models that biologists of the era used – cutting up and characterising and classifying all of the planet's creatures and their constituent parts – was proving hopelessly inadequate.

When Delbrück saw phages at work for the first time in 1937, he was delighted – this ultra-basic form of life, which appeared as spots on a dish within a few hours, was exactly what he had been looking for in his quest to understand the fundamental processes that drive the living world. The 'atoms of biology', he called them.[3]

As the world descended into war, his work inspired a trio of phage enthusiasts to come together, against the odds, from countries that were now sworn enemies, forming the nucleus of what would become known as the Phage Group. Together, these 'two aliens and a misfit', as they referred to themselves – with the help of a few very special phages – completely revolutionised the way scientists study life.

Max Delbrück was born in 1906, on the western outskirts of Berlin, into the sort of family that assigned him a number at birth: 2157, meaning he was the seventh child of the fifth child of the first child (and so on) of Gottlieb Delbrück, a family patriarch who lived in the eighteenth century. His elderly father was a professor of the art of war at Berlin University, and his extended family contained all sorts of famous and learned figures, including a government minister and a chief justice of Germany's Imperial Supreme Court. As a young child, when a family friend remarked that famous people don't have famous children, he apparently shot back: 'Just wait!'[4]

By 1918, when Delbrück was twelve, three-quarters of the young men in his family had been killed in World War I,[5] and his home suburb reduced to an eerie ghost town. But the horrors of these years did not stop Delbrück from following his ambitions, deliberately turning his interests to subjects his high-achieving family and their learned friends knew little about, such as mathematics and astronomy.[6] His loud alarm clock would often wake the family when he needed to see some distant cosmic pageant with his two-inch telescope in the middle of the night. He grew into a very tall, very thin, eternally youthful looking man, with a big square forehead, horn-rimmed glasses and thick tufty hair that together screamed 'theoretical physicist'.

At seventeen, Delbrück left home to study the then-blooming science of astrophysics, moving universities twice before finally settling on Göttingen. He arrived in 1926 as some of the biggest names in physics discussed ground-breaking theories on the mechanics of the universe; he recalls shuffling into a seminar behind Albert Einstein to hear Werner Heisenberg's first explanation of his theory of quantum physics.[7]

As Adolf Hitler, and the National Socialists, began their rise to power in Germany, Delbrück began to hold regular meetings with five or six eminent physicists at home, where they would talk long into the night and could share ideas away from the watchful eye of the German authorities. The group expanded at Delbrück's request to include notable biochemists and biologists, too. 'Two or three times a week we met . . . we talked for ten hours or more without a break, taking some food during the session,' wrote the biologist Carl Zimmer.

It was in these extra-curricular meetings that Delbrück began to tune in to the grand unanswered questions in the science of life. Having struggled through his physics PhD, which he recalled later as 'dull' and 'a nightmare' because he had failed to develop any kind of big new idea,[8] Delbrück wrote a speculative paper that suggested each gene might be encoded by a chemical, and that the properties of the atoms or molecules within it might be related to the eventual function of the gene. Just like in quantum physics, where a whole new set of rules had been devised to understand how matter works at the subatomic level, Delbrück also thought that a new set of laws might be needed to explain how molecules work in a system that is alive.[9]

The paper caught the attention of some of the biggest names in physics, including Erwin Schrödinger – he of the famous dead-and-alive-cat-in-a-box thought experiment – who also wanted to try to understand the physical laws of living creatures, and why life was so different to the way the rest of the universe worked. Schrödinger cited Delbrück's idea in his seminal book, *What is Life?*[10]

Delbrück took positions in Bristol, where he learned English, and in Copenhagen, where he flourished under the remarkable physicist Niels Bohr, but the build-up to World War II would force him to rethink his academic ambitions in Germany. As his Jewish colleagues fled persecution, his own career was stalling: he had failed to show sufficient 'maturity' at a series of Nazi *Dozentenakademie*, or lecturer's indoctrination camps, following a move back to Berlin. Academics had to pass tests at such camps before being granted permission to teach or become lecturers, and the extremely direct Delbrück had not only refused to play ball at his first camp, but also at a second. By his third, Delbrück recalls: 'Everybody knew by then what you could say and couldn't say, and everything was much more relaxed. But still I must have shot my mouth off.'[11]

As well as regular trips to the *Dozentenakademies*, Delbrück was forced to go to increasingly arduous lengths to prove he wasn't Jewish.[12] In 1937, when the American philanthropic organisation the Rockefeller Foundation offered him a fellowship to travel to the US to expand on his ideas about genes at Caltech, Delbrück took his chance to escape the increasingly sinister atmosphere at home. It would be Delbrück's first position as a fully fledged biologist.

His first forays into genetics in the US were frustrating and disappointing. Following the lead of all good geneticists of the era, he began studying the genetics of the fruit fly *Drosophila*,[13] a supposedly simple biological system for conducting experiments. However, even this comma-sized insect is an extraordinarily complex thing – a mess of delicate intermingling tissues and hundreds of types of cells. Delbrück found the terminology of the various genes, body parts and mutant breeds – many of which were given weird names – unending, impenetrable and impossible. An astronomer and theoretician at heart, used to working with numbers and equations, not unpredictable little creatures with silly nicknames – he needed something far simpler, more elemental. On top of the interminable terminology of fruit fly biology, as a non-Jewish German in the US, he found he was often suspected of being an undercover Nazi agent.[14] He was miserable.

It had been a camping trip organised by a colleague that caused Delbrück to miss the seminar on bacteriophages given by Emory Ellis. Even in the late 1930s, hardly anything was known about viruses – what they really were and how they replicate. A few years earlier, in 1935, work by a young scientist called Wendell Stanley had suggested that viruses were made of protein, and with many scientists beginning to think genes were proteins too, ideas about what genes, proteins and viruses were had started to bleed into one another. The science of bacterial viruses was even more confused. The debate over the 'd'Herelle–Twort phenomenon' was still going, decades after d'Herelle's discoveries.

Delbrück had 'only vaguely' heard of phages before deciding to seek out Ellis to hear what he'd missed.[15] But he immediately saw something in these viruses that others didn't: a fast and simple tool with which to study reproduction, a defining feature of life, at its most elemental – and with only simple lab equipment. He thought the simplicity of a virus might allow him to work on the fundamental laws and physics of life and reproduction without the complexity of a living organism around it. 'I was absolutely overwhelmed that there were such very simple procedures with which you could visualise individual virus particles,' recalled Delbrück. 'I mean you could put them on a plate with a lawn of bacteria, and the next morning every virus particle would have eaten a 1mm hole in the lawn.'[16]

Delbrück immediately asked if he could join Ellis in his work, learned the rudimentary techniques of 'phagology', and fell in love with the viruses of bacteria. Not, initially, because of some deep interest in these tiny beings – in fact, he initially joined the camp that believed phages were proteins, not microbes – but because he could set up an experimental question and have data and answers the next day.[17] And with such basic equipment too! He could throw out his books on fly genetics, and he did.

Together, Delbrück and Ellis developed what was known as the one-step growth experiment, where a batch of phages and host cells were allowed to go through only a single life cycle of infection and replication, under carefully monitored conditions.

By measuring the changing concentration of phages at various points in the experiment, researchers could plot graphs revealing the intricacies of the phage life cycle – how long it took for the phages to replicate inside the host cell, known as its latent period, and how many new viruses were liberated when it burst, known as its burst size. Delbrück and Ellis's one-step growth curves would become central to many studies of bacteriophage biology for decades to come.

After working together for a few years, Ellis changed the focus of his research completely, returning to work on more direct questions about cancer. The dispute over the nature of phages still hung over the field like a warning sign not to enter, leaving Delbrück virtually alone in his interest in the bacteriophage as something worth studying. But thankfully, he would not be alone for long.

Around the same time that Max Delbrück was finding his passion for biology, in Italy a young doctor called Salvador Luria was growing bored of medicine. Born in Turin in 1912 to a Jewish family, Luria had specialised in radiology after medical school, then spent time as a medical officer in the Italian army. The young Italian had only gone into medicine because his parents wanted him to. He later described life as a trainee doctor as 'drudgery' and felt ill-prepared for the inevitable emergencies he might encounter in practice. He said his chosen field, radiology, 'turned out to be the dullest of medical specialties'. Despite his shortcomings in science at school, he dreamed of being a theoretical scientist, like the famous European physicists developing mind-boggling new ways to describe the universe.

While working as a radiologist in Rome he began teaching himself calculus and physics in his spare time, and also tried to understand the complex genetics of fruit flies. And he read with awe about Delbrück's idea, born out of a little dinner club in Berlin, that the functions of living matter might be controlled by some unknown characteristics within the molecules of genes. He began to wonder how one might test such an idea. To his eternal gratitude, Rome's unreliable transport system helped provide an answer. The city's electric trams often broke down due to power failures, and when one day he found himself stuck again in a stalled car, he decided to strike up a conversation with a man he recognised. That man turned out to be Geo Rita, professor of virology at Rome University. As they chatted, Rita told Luria all about his work sampling the bacteria of the Tiber River – and the viruses that had been found living alongside them.

In the days after their impromptu conversation, Luria began to wonder if the virus of a bacterium might be such a basic form of life that it could help scientists understand biology's most fundamental processes. This was only months after Delbrück's epiphany in Emory Ellis's lab. When Luria discovered that Delbrück was also working on bacteriophages, his excitement grew. He'd had the same idea as one of the world's leading thinkers in a fascinating new frontier of biology.

Luria – small, smartly-dressed and with an almost cartoonish friendly face – had also found himself working in a country darkening under fascism. Hoping to emulate Delbrück and escape Europe, Luria applied for a fellowship to study in the US and was accepted. But the very next day, 18 July 1938, Mussolini's government announced its *Manifesto della razza* – a new racist charter aligning the country ideologically with Nazi Germany. These grim anti-Semitic laws stripped Jews of their citizenship and expelled them from professional positions. His chances of a fellowship vanished overnight – 'like an ephemeral flower', as Luria put it.

Instead, Luria moved to Paris, alone and with no money. He was taken on by the renowned physicist Fernand Holweck at the Marie Curie Radium Institute, who liked Luria's combination of ideas in medicine, radiation and physics, and who let him work for the first time in a lab on his own. Luria set to work on his new favourite life form, bombarding phages with radiation to try and understand their molecular make-up.

Within a few months, however, war had broken out in Europe. In his autobiography, Luria remembered that he was, bizarrely, 'cycling around Brittany on a borrowed bicycle when World War II started'. As Hitler's forces approached the French capital in May 1940, and Parisians fled the city in their thousands, Luria took a final walk through the eerily empty streets before fleeing by bike.

'For a couple of days I managed to keep just ahead of the German troops,' he recalls casually. 'I was twice the target of ineffectual strafing from planes . . .'[18] For a month he travelled around France escaping the warfare by bicycle or by jumping freight trains. The goal of his 500-mile 'peregrination'* was the relocated US Embassy in Marseille, on France's southern coast, and the possibility of emigration to the other side of the Atlantic.

As Europe began to explode into violence, the path of Luria's life – and the fate of phage science – lay in the hands of a faceless bureaucrat and their stamp. Scores of hopeful refugees – from shabby peasants to famous intellectuals – waited in the US consulate, where just one or two cases were processed each day. Luria managed to out-bureaucrat the Embassy bureaucrats, obtaining approval for a US visa via a ridiculous route that involved him getting several transit visas for Spain, Portugal and the Belgian Congo. Eventually, his US visa was approved, and the echo of that stamp still reverberates through science to this day.

Luria made his way out of occupied France and through Spain to the coast of Portugal, and in September 1940 finally boarded the S.S. *Nea Hellas* for the US. Just to ensure his escape from Europe didn't go too smoothly, the S.S. *Nea Hellas* crashed as it entered New York Harbour, gouging a deep cut in a Norwegian freighter. But Luria had made it. And he immediately tried to contact Max Delbrück.

Fellow emigrant scientists helped Luria into a research post at Columbia University, in New York, and at a conference in Philadelphia in late December that year, he finally met the one other person in the world who thought phage was the key to

---

* The only other time I have encountered this strange word in print, meaning a long meandering journey, is in the title of Felix d'Herelle's memoirs: *Les pérégrinations d'un microbiologiste*, or 'The wanderings of a microbiologist'.

understanding the mysteries of genetics. He and Delbrück not only agreed to join forces then and there over dinner, but like new lovers in the throes of new passion, immediately travelled to New York on New Year's Day 1941 for a 'forty-eight-hour bout of experimentation' in Luria's lab. Using dishes of bacterial viruses as their biological looking glass, the pair of immigrant phageophiles began to seek answers to some of the most intractable questions of the century.

With Luria at his side, Delbrück pressed on with his quest to answer the most vexing questions in biology – What do genes consist of? How are they passed onto offspring? How do they act in a living cell to create characteristics? He still doggedly believed that some new law of living physics was round the corner.

In 1943, he and Luria made a huge breakthrough in understanding bacterial genetics that would have been the highlight of most scientists' careers. After exposing multiple batches of identical bacteria to phages and culturing each one for several generations, they observed great variation in which bacteria developed resistance and when. Some bacterial plates were full of resistant cells, some contained just a few, others none. The development of resistance was so irregular that it seemed random – and indeed, they were able to prove mathematically that any helpful genetic mutations that gave the bacteria resistance *were* arising at random, rather than as a response to the addition of phages.

The work showed that even organisms as simple as bacteria evolved in accordance with Darwin's theory of natural selection, and provided the first strong evidence that bacteria had genes, just like more complex organisms. The methods they developed gave rise to an entirely new and exciting field of science – 'bacterial genetics'. Bacteria and their viruses would eventually become the go-to lab organisms for studying genetic changes, evolutionary change, and the complexities of life. Delbrück and Luria's findings also helped put to bed the idea that organisms can acquire the genetic adaptations they need during their lifetimes, a competing evolutionary theory to Darwinism known as 'Lamarckism' which had stubbornly refused to die.

This was massive news in biology, and for Luria it was one of the great highlights of his career. But with his eyes on some grand unifying theory of biology, Delbrück described these discoveries as nothing but a 'side issue'[19] which threatened to displace his focus on the big question: *what is life*? And so he and Luria continued with their original inquiries. Delbrück focused on the question of reproduction, Luria the nature of the gene. They made many more wonderful discoveries about the nature of bacteria and viruses almost as a byproduct of their investigations.

By this time Delbrück and Luria were also in communication with an American microbiologist called Alfred Hershey. Unlike Delbrück and Luria, Hershey had actually begun his career studying bacteria, but had been working with phages for many years. In his previous roles he had been tasked with helping older scientists prove the incorrect idea that phages were some kind of protein or enzyme, not a living structure[20] – fruitless work that had made him miserable. Delbrück admired the precise use of mathematics in his papers and promised if Hershey joined the collaborations between himself and Luria it would be more rewarding. 'You have spent too long in the depressing atmosphere of piddling bacteriologists,' he wrote.[21]

When they finally met and began working together, Delbrück's first impressions of Hershey – a small and serious-looking man whose round glasses and thin moustache remained unchanged his entire career – were of a man who liked to be alone. 'Drinks whiskey but not tea. Simple and to the point. Likes living on a sailboat for three months . . .'[22] Colleagues would later write of him that in social situations, he would often reply to questions with a straightforward 'yes' or 'no', and in scientific meetings, one might get no answer at all, which was his economical way of saying that he had no thoughts on that subject at that time.[23]

The three men made an odd trio, and it was Hershey who famously referred to their group as 'two enemy aliens and a misfit'.[24] Many traditional geneticists remained sceptical that their approach would yield any further insights.[25] But Hershey added yet more experimental creativity and clout to the great mind of Delbrück and the sponge-like brain of Luria. Gradually, more and more scientists came from around the world to join the eccentric trio in their pursuit of clues from bacteria and viruses about reproduction, genes and – perhaps – life itself. It was the start of a genuine revolution in the science of biology, our understanding of nature and our ability to manipulate and change it. The ever-expanding group of scientists using phages to understand life became known as the 'Phage Group', and sometimes the 'Phage Church'.* Yet even as

---

* The American molecular biologist and Phage Group member Frank Stahl wrote that the Phage Church was 'led by the Trinity of Delbrück, Luria, and Hershey . . . Delbrück's status as founder and his *ex cathedra* manner made him the pope, of course.'

this exciting new group became experts in using phages as tools for scientific reasoning, the nature of the viruses themselves remained a mystery.

On the other side of the Atlantic Ocean, another German physicist called Ernst Ruska had been building a device with the potential to end the ongoing debates about what viruses were with just a few adjustments of its brass dials and clunks of its stiff steel switches. Since the 1920s, Ruska had been quietly working on the idea that a stream of electrons, the subatomic particles that normally orbit the nucleus of atoms, could be focused through a special lens into a beam and used to explore matter that was too small to see with a light microscope. By 1931, he had developed a strange-looking device which looked rather like an elaborate shisha pipe sitting on top of an old gas oven. It allowed him to focus a beam of electrons on a very thin sample and create an image from what passed through. Within a couple of years, he was able to make images of objects far smaller than had ever been seen by human eyes before. The electron microscope was born.

Twort, d'Herelle and Delbrück had all leaned into their work on phages fully accepting that whatever the nature of this stuff was, it was far too small to ever be seen or investigated by direct observation. They used their scientific creativity to devise experiments that summoned visible effects from this hidden world.

Sometimes this was just about sheer numbers: a flask thick with billions of live bacteria is opaque and milky; when it goes clear, we know that the bacteria have been decimated. The return of the clouding, days later, heralds the growth of phage-resistant bacteria, a proliferation of the tiny number that were somehow naturally resistant to the phages and have grown into a resurgent army.

Other times it was about maths and logic – add viruses with known characteristics to hosts with a range of precisely bred traits, note the pattern of what grows and what doesn't and then analyse the heck out of what happened and what it means.

Delbrück and Luria continued to work happily and product-ively in the darkness – they knew no other way. But in the late 1930s, while Ernst Ruska worked on refining and patenting his invention (initially marketed as 'the hypermicroscope'),[26] his brother Helmut, a biologist and doctor, started to look at biological specimens with the powerful new machine. Near the top of his long list of things to investigate was the mysteri-ous bacteriophage.

It's unclear what Delbrück or d'Herelle thought phages might 'look' like – or if they had even considered the question at all – but it is fair to say they would not have guessed what Helmut and Ernst saw when they first focused their electron beam on a super thin plate covered in phages. The images, first published in 1940, were grainy and blurry, but remarkable. They show round phages, club-like phages, phages with heads and tails, some surrounding their bacterial prey, others spilling out from broken cells, whole tableaus of microbial warfare magnified by 25,000 times their natural size. But the Ruskas were based in Germany, and Helmut's findings were published in a German journal; the sanctions and chaos of a world at war prevented the communication of these historic images to the people who wanted to see them the most.

The phage remained an abstract, divisive concept well into the 1940s. Meanwhile the debate over the nature of phages continued to cast a shadow over the subject and Delbrück's mission. The effects seen in flasks and on dishes were just that – effects, phenomena, numbers on graphs, arrows on blackboards. Without those grainy images, there was little solid evidence about what the phage really *was*, what structure it had, how it looked. There were still many in the scientific community that held the view that phages – or 'the Twort–d'Herelle phenomenon' to those of the old school – were not a distinct form of life at all, but self-destructive proteins produced, for some reason, by the bacteria themselves. Even Delbrück himself referred to viruses as 'large protein molecules' in a paper as late as 1939.[27]

Eventually, electron microscopes began to be installed in labs outside of Germany. In 1942, the biologist Thomas F. Anderson – having at first believed the electron microscope was a Nazi hoax – managed to get funding to explore biological specimens with one being built by scientists at the Radio Corporation of America (RCA). Soon many were queuing up to put very tiny things into this very expensive photo booth. While Delbrück was happy to use theory and maths to dredge up information from the darkness, Luria knew that if they were to make any real progress, they needed to know exactly what a phage was. So, he reached out to Anderson and asked to collaborate on an investigation of phage with the electron microscope.

To get a decent image of the viruses, Luria had to produce a solution of phages concentrated enough to ensure that some would be caught in the fine beam of electrons fired through the device. After failing to produce a solution strong enough the first time, Luria's second attempt worked. With a solution they calculated contained tens of billions of phages in every millilitre, Anderson helped produce a series of black and white electron micrographs[28] that gave the burgeoning Phage Group the first glimpse of their precious babies: not one but dozens of them, caught in the act of surrounding and attacking a great pill-shaped *E.coli* cell.*

It's hard to imagine the absolute astonishment that all microbiologists would have felt at seeing dozens of distinct, circular or lollipop-shaped forms† floating around and attacking the walls of bacterial cells. The notion that viruses were enzymes, ferments, secretions or, going back further, 'poisonous liquids', had finally been conclusively disproven.

On seeing the images, Delbrück is said to have uttered, 'I don't believe it!' again. He, Luria and Anderson took a series of further pictures to capture important points in the replication cycle. They were able to see that the phages were indeed multiplying inside the bacterial cell, which then ruptures and releases more – a process also predicted by d'Herelle through his experimentation.

The old master was apparently on his deathbed when he first saw an electron microscope image of a phage, years later in 1949.[29] Over thirty years since he first deduced that what

---

* The phage scientist J. J. Bronfenbrenner, one of the first scientists in the US to study phages after their discovery in Europe, and a staunch believer that they were more likely to be a complex molecule than a living entity, is reported to have cried out when he saw the images: '*Mein gott*, they've got tails!'

† Most descriptions of these first images describe the phages as 'tadpole-shaped' or 'sperm-shaped', but I think the 'tails' of the virus look rather stiff and straight, like the stick of a lollipop.

he had seen must be a form of replicating life – a microbe of a microbe – and after two decades of vitriolic arguments, pointless polemics and personal smears – here was something like proof. The phage was a tiny but distinct biological structure, many of them with both a head and a tail and even spider-like leg structures – far more interesting than the blobby bacteria, and far from simply a destructive enzyme or bacterial 'ferment'. But in the final few last years of his life, d'Herelle had seemingly grown tired of the old debates and was writing long meandering essays full of his dogmatic views on science, politics and life. There were some brief moments of overlap between d'Herelle and the new masters of phage science – Delbrück and Luria, at one point in the 1940s, presenting their latest work to d'Herelle's protégés in France – but the founding father of phage science on the whole is said to have paid little interest to the exciting new role the organisms he discovered were playing in a scientific revolution.

The electron microscope also showed the Phage Group something fascinating: that the infecting phages never entered the bacterial cell; instead they latched on to the outside, and stayed there, until the cell mysteriously burst forth with many dozens more half an hour later. For Delbrück – at heart a physicist desperate to find some simple maths-like truth at the heart of biology – this was just frustrating: the replication of even a simple thing like a phage was obviously much more complicated than he first thought. 'Any living cell carries with it the experience of a billion years of experimentation by its ancestors,' he remarked later. 'You cannot expect to explain so wise an old bird in a few simple words.'[30]

But for the others, the images sparked their imaginations.

The sperm-like appearance of the particles, and the fact that they bind to the outside surface like a sperm to an egg, led some to theorise that rather than being a parasite of the bacteria, the phage was in fact somehow fertilising it.[31] Meanwhile Anderson, the microscope operator, looked at the 'ghosts' – the empty-headed phages stuck to the outside of bacteria – and expressed a 'wildly comical possibility.'[32] Perhaps the phages injected some kind of genetic material, syringe like, into the host cell, and it was this that changed the host's nature from thriving cell to suicidal virus factory? Around this time, other work by members of the Phage Group had determined that phages contained just protein and DNA, nothing more. And so, one of these must be the genetic material that carries the blueprint of life. Together, they were getting closer.

In 1945, as World War II finally came to an end, Delbrück organised a summer course on his pioneering phage work at Cold Spring Harbor in Long Island. For scientists working in this exciting field in a newly optimistic post-war world, Cold Spring Harbor was like summer camp. Small purpose-built labs and charming lecture halls sat nestled like holiday homes around a glorious bay, sheltered on all sides by forested hills and with its own private beach. It was not only a beautiful place to work but a perfect spot for swimming, fishing and watersports in summer, the lush hillsides turning a thousand shades of red, green and gold as autumn came.

The Danish biologist and student of Delbrück, Niels K. Jerne writes – in a wonderfully melodramatic way – of the atmosphere among this small but influential group of scientists as the 1940s became the 1950s:

> Nature winced under the onslaught of young men counting specks in Petri dishes ... The air was filled with the phage particles that Delbrück had picked out as one of the weakest spots in the armour behind which nature guards her secrets. Over it all hovered the spirit of Delbrück, who was shepherding his handpicked band along the last stretch of the narrow path to the central fortress of biology . . .[33]

Jerne remembers scientific 'jam-sessions' with Delbrück, 'attired as if we had just been expelled from a tennis court'. Others remember hot, lazy nights hanging out by the waterfront, or drinking beer in the nearby village, and endless practical jokes – from letting the tyres down on friends' cars to soaking their beds in ice water. One scientist's attempt to hold a staid evening event was attacked by scientists bearing toy machine guns. Sepia pictures show Delbrück and Luria reclining on sunny porches in short-sleeved shirts or shirtless, looking more like two old GIs having a beer rather than the great intellectual powerhouses of the era discussing one-step growth equations.

But Delbrück was also a fearsome leader of this growing movement. He only accepted students who excelled in the fiendish maths required to do the precision biology he was developing and would discriminate between scientists who had and those who hadn't taken his lakeside course. He began referring to the old methods of biology – the endless observation, naming and cataloguing of living organisms, as 'stamp-collecting'.[34] He could be brutal in the way he challenged his students and peers to produce original and robust scientific ideas. It was not only common for him to shout 'I don't believe a word of it!' to colleagues when unconvinced by their theories, but he would often interrupt seminars to say they were the worst thing he'd ever heard, or just walk out in the middle of them. The American biophysicist George Feher once remarked that he was pleased to be giving *two* talks in a series of seminars with Delbrück in San Diego, 'as at least one of them wouldn't be the worst thing Max has ever heard'.[35]

Under this demanding and exacting environment, a golden age of biological research began, with Delbrück's highly skilled group of protégés expanding the Phage Group and developing a new form of science seldom seen before in biology. The group used scintillating logic to devise variations of virus–host experiments that would reveal the inner workings of life in the form of mathematical data. Suddenly they were making leaps and bounds in the understanding of how viruses replicated within their host and developing a more precise understanding of genes and genetics. With d'Herelle's reputation now shot, and phage therapy increasingly seen as archaic and obsolete in the West, Delbrück began to drag phage science out of its amateurish era. Anderson, the biologist who produced those first phage images using the electron microscope for Delbrück, recalled how scientific papers carrying the latter's name 'formed a little green island of logic in the mud-flat of conflicting reports, groundless speculations, and heated but pointless polemics that surrounded the Twort–d'Herelle phenomenon.'[36] The phage courses at Cold Spring Harbor marked 'the onset of biology as an exact science', wrote Delbrück's biographer and last graduate student, Ernst Peter Fischer. The courses would run for almost twenty-five years and produce a distinguished family tree of biologists who would fan out into many areas of life science and make countless wonderful discoveries. 'All of a sudden, the whole world begins to take an interest in my viruses,' Delbrück wrote as scientists from around the world came to Cold Spring Harbor to study.

This expanded Phage Group eventually helped solve the mystery of how phages can lurk inside bacterial cells for generations without producing any progeny, an issue that had been causing much confusion and causing many to refute the idea, still, that the phage was a virus. The French scientist André Lwoff, who had worked for years at the Pasteur Institute before joining Delbrück at Caltech, had made it his personal mission to demonstrate the importance of these questions, after the husband and wife scientists who had been studying it since d'Herelle's time – Eugene and Elisabeth Wollman – died in a concentration camp during World War II.[37]

Lwoff, unlike Delbrück, preferred physical experiments to mathematics, and simply took individual bacterial cells, exposed them to different conditions, and observed them carefully under the electron microscope. Using the relatively massive bacteria *Bacillus megaterium*, he was able to directly observe that bacteria infected with phages could happily live and divide as normal without producing new viruses; and that if they were forcibly split open, they contained no new viruses; but that they also could, somewhat randomly, start producing new viruses. It was not yet clear that it was phage *DNA* carrying this potential to produce viruses. But Lwoff had deduced that an invisible essence of the virus was all that was needed to make a cell produce phages at some point in the future. Further experiments by Lwoff revealed that stressing cells, for example with UV radiation, caused the phage to switch mood, from an unobtrusive guest to one suddenly demanding the production of new phages and bursting its host.[38]

A year before his courses started, Delbrück had devised and negotiated what became known as the Phage Treaty. It was an agreement that goes a little way to explaining why we know so little about so many of the millions or billions of different types of phages out in the world. Scientists from institutions around the world agreed to work only on seven distinct phages, named T1 to T7, and to work with the same nutrient broth, at the same temperature, 37°C. All seven phages produced plaques reliably and consistently in experiments, and preyed on *E.coli* bacteria, which was quickly becoming established as the work-horse bacterial species in microbiology and genetics because of the ease with which it could be grown and studied in the laboratory.

Up until then, biologists studying phages in different labs all had their own collection of phages on which they worked, which made it difficult to compare results or build a deep body of knowledge on how one particular virus and its host interact. The Phage Treaty undoubtedly helped accelerate progress in the core biology of phages, plus it consolidated Delbrück's position as the supreme leader of the phage world: if you wanted your work to be noticed by the great Max Delbrück, you had to work with the right phages.

It also meant that the growing network of phage scientists around the world concentrated their focus on just a tiny set of the planet's phages. Of the seven, the so-called T-even phages (T2, T4 and T6) turned out to be most useful for biochemical and genetic studies. All are believed to have first been isolated from sewage or poo and, indeed, these useful little scientific tools are present in most animal and human guts.[39] Delbrück's treaty once again helped propel the science of phages forward, with comparable data from many research groups around the world helping to build a detailed picture of the fundamental systems that drive cellular and non-cellular life. But, the fact remained, the true abundance and diversity of phages out there in the world remained unseen and unnoticed.

For all his wonderful talents as a scientist, Delbrück did have blind spots, and his dogmatic and stubborn approach to science and how it should be done was one of them. Not unlike d'Herelle, he had immovable views about the right way to do things — dismissing methods that he didn't feel were precise enough, refusing to look at data from work done on phage strains from outside his treaty and railing against the academies and governments that funded his work. He was particularly dismissive of chemistry.

His biographer, Ernst Peter Fischer, who was also Delbrück's last student and who interviewed him at length shortly before his death in 1981,[40] confirmed this to me. 'He hated things he didn't know or understand,' Fischer observed.*

Fischer recalls him and other physicists joking that a recently separated colleague was less upset about his devastating divorce

---

* It is, still, not uncommon for both biologists and physicists to look down on chemistry somewhat as a subject, with its lack of 'big ideas' or existential theories and its smelly, fumy experiments. Chemists would argue all of biology is simply chemistry's most outrageous display of what it can do.

than the fact that his wife had left him for a *chemist*. Delbrück's scientific snobbiness would come back to bite him as other scientists, including his own peers and protégés, inched closer to the great biological breakthrough of the twentieth century (and arguably, of all time).

By the 1950s, as the Phage Group found even the replication of a virus was more complex than they could ever imagine, biochemists were starting to make important strides in narrowing down the chemicals known to be important in replication and inheritance.[41] There was increasing evidence from the biochemists that a chemical called DNA was at the heart of it all. Hershey, alongside his brilliant research assistant Martha Chase, had devised an ingenious experiment that proved DNA was the material that entered the bacterial cell from the phage. Known as the 'blender experiment',[42] the pair had 'labelled' the protein and DNA in a sample of phages with different radioactive isotopes, so they could track and trace where those particular molecules had gone when the phage infected a bacterium. After trying a range of high-tech techniques to shake the phage 'ghosts' off the bacterial cells so they could be analysed separately, they eventually borrowed a colleague's kitchen blender to tear the tiny beings apart.[43] It had just the right amount of shearing force to rip the phage's binding sites away from the bacterial cell walls but not so much that it chopped them all into pieces. Now able to separate off the phages and the bacteria, they found the radioactive DNA had moved from the phage into the bacteria. The radioactive protein, however, was still in the phages. The improvised experiment showed clearly that phages' DNA was transferred into the bacteria during infection and nothing else.

Hershey and Chase's virus smoothie strongly suggested that it was in fact DNA not protein that carried chemical instructions into the bacterial host. But Delbrück did not believe a word of it. He wasn't interested in complex chemical experiments – he wanted to theorise his way to a solution. He even once said that he believed DNA was a 'stupid molecule', incapable of forming complex structures.[44]

'For Delbrück, chemistry had no grandeur, no theory,' Fischer tells me, laughing. 'And of course, he hadn't learnt it, either.'

Other members of the Phage Group were less dogmatic though. Luria was more open to the idea that understanding the chemical structure of genetic material might unlock the secret of how it works. Knowing he could never learn the intricacies of biochemistry himself, he agreed to let one of his students go to Cambridge to study with a biochemist and learn the tools of the trade. That student, James Watson, is now maybe as famous for making racist comments and for his misogynistic views of colleague Rosalind Franklin, as he is for his contribution to one of the most famous scientific papers in history.

Watson started his career as an unknown, ambitious and rather arrogant biology graduate who – in his own words – had 'managed to avoid taking any chemistry or physics courses that looked even medium difficulty' owing to his laziness.[45] After obtaining a PhD on phage genetics at just twenty-two with the Phage Group, he was sent away by Luria on what was initially meant to be a short fellowship to learn more about general biochemistry. But after travelling Europe, he began to indulge his new interests: using X-rays to interrogate the atomic structure of interesting molecules like DNA.

At Cambridge, he met Francis Crick, who agreed that understanding the 3D-chemical structure of DNA was surely a massive part of the puzzle of life. The pair got along famously (literally) and started to formulate ideas on how to do it. Watson ended up repeatedly extending his stay in Cambridge as the pair got closer and closer to understanding the structure of DNA and, with it, the whole secret of how information is encoded and replicated in cells.

As Watson – alongside Crick, Franklin, Maurice Wilkins and Linus Pauling – honed in on what would become the most exciting discovery of the century, Delbrück remained unimpressed. One time, when a colleague mentioned Watson's love for the work he was doing at Cambridge, Delbrück apparently threw his hands up in disgust.[46] When the pair finally met at a conference to discuss the work, Watson, still technically a postdoctoral student of Luria's, recalled that Delbrück 'seemed bored talking about it'. 'Even my information that a pretty X-ray picture of DNA existed elicited no real response . . . In Delbrück's world no chemical thought matched the power of a genetic cross.'[47]

By 1952, Watson and Crick were on the verge of announcing they had solved the puzzle of DNA's structure. The towering model they built in their lab from clamps, tin and cardboard not only revealed what DNA's molecular structure looked like – the famous double helix – but also how the miraculous chemical worked: how the different subunits of DNA, one after the other, formed a kind of long chemical code. The double-stranded structure also immediately revealed the elegantly simple way the whole thing can replicate itself, simply unzipping along its length and then repairing itself with complementary subunits, each side of the zip forming two new strands.

Some of the most significant letters in the history of science pinged slowly back and forth across the Atlantic as Watson kept his mentors updated, despite their apparent lack of interest. Finally, Watson sent a letter to Delbrück explaining what they had discovered – the structure of DNA and with it, the secret of what genes are made of and how life passes information from one generation to another. 'That's it, that's the end of it,' Delbrück apparently said when he read the letter and saw all the bits of the great puzzle fall into place. In one blow, the double helix revealed the 'secret of life'* – and there was no great paradox, no new rules of physics required to explain it. It was just chemistry; chemistry so simple to understand that it, in Delbrück's words, 'made the whole business of replication look like a child's toy that you could buy at the dime store'.[48]

The paper Watson and Crick published to huge interest in 1953 revealed to the world that now-familiar twisting shape of the DNA molecule, and the pair went on to international fame and fortune. The photos of them next to their tabletop model became the defining image of the start of the genetic revolution. Our beloved phages, having contributed so much to the development of genetics and molecular biology, were buried under the weight of a simpler scientific story – that Watson and Crick deduced the secret of life with the help of Rosalind Franklin's famous 'photo 51'.

Over the next fifteen years, a vast array of molecular

---

* According to legend, Watson and Crick ran into *The Eagle* pub in Cambridge and Crick declared to drinkers that they had 'found the secret of life!' However, Crick has no memory of this, and many other aspects of Watson's version of events have proven to be unreliable.

biologists – often using phages – revealed how the chemical code of DNA is 'translated' by chemicals in the cell into proteins with different functions, which in turn form the tiny signals and structures that build all living organisms, including us. Delbrück, however, was not among them. After continuing to try and understand replication in more detail for a time, he never published anything on phages or genetics again, instead moving on to another problem entirely, studying how the cells of the simple fungus *Phycomyces* respond to stimuli to try and better understand more complex sensory systems.

Delbrück had discovered the hard way that phages were not the 'naked genes' he thought could reveal nature's secret to him. They were highly specialised, deceptively complex biological entities that do a lot with very little. But while he may have failed in his goal to discover the beautiful process that drives life on Earth, the legacy of his work has been immense. His vision of the bacteriophage as a system capable of revealing precise answers to biological questions created an experimental tool which continued to change science and the world long after he had lost interest in them, long after he died in 1981, and will continue to for a long time into the future. He, Luria and Hershey had essentially invented an entirely new field of the life sciences that would become known as molecular biology, which, in turn, would lead to genetic engineering, genetic medicine and a host of ever-more fantastical biotechnologies that are now changing the possibilities of biology at an unprecedented pace. As a model organism, phages' contribution over this period is perhaps unrivalled in the history of science (although the countless mice, worms and flies that have died in the name of science might have something to say about that).

By my calculation they have been central to at least six Nobel Prizes, shared by fourteen Nobel Laureates and indirectly helped a further fifteen scientists make the breakthroughs that won them the famous Swedish medal.[49]

Work on phages led to the understanding that both viruses and bacteria had genes, opening up a clear, clean canvas on which genetics could be explored precisely and in fine detail for the rest of the century and beyond. After Hershey and Chase helped provide experimental proof that DNA, not protein, was the genetic material that drove life, phages then helped scientists reveal how cells translate genes into proteins, turn genes on and off, and how the molecules of different proteins fold into 3D-shapes that give them function.

They helped reveal the concept of 'horizontal gene transfer', the mechanism by which bacteria of unrelated species can swap genes, helping to refine our understanding of evolution. The first gene to be sequenced by scientists – that is, the exact sequence of DNA subunits in a gene deciphered by scientists – was a gene from bacteriophage MS2. The first entire genome to be sequenced – in other words the first entire set of genes that make up an organism to be decoded – was also a phage (the super simple bacteriophage ΦX174). The first synthetic organism, made from DNA synthesised in a lab rather than grown naturally? Also a phage, based on the super minimal genome of ΦX174.

Phage researchers, studying the basics of DNA mutation and repair, provided the basis for our contemporary understanding of cancer, and researchers studying how phages infect their host led to an understanding of how human viruses infect and spread in our cells and cause disease. And all that before we even get to genetic engineering.

Once a controversial technology that people feared allowed us to 'play God' with nature, genetic engineering is now used every day to reshape living systems for our needs in research, agriculture, medicine, biofuel production and even conservation. It's all thanks to phages. Genetic engineering would be impossible without special DNA-cutting enzymes called restriction enzymes, discovered by studying how bacteria defend themselves from phages. These 'DNA scissors' also led to the development of DNA fingerprinting (also known as DNA profiling), the forensic technique that has helped solve millions of criminal cases and has probably prevented many more since its invention in the mid-1980s. Enzymes that scientists use to stick chunks of DNA together – key lab tools known as ligases – were first isolated from the genome of the T4 phage.

Trace almost anything being done in a modern molecular biology lab back to its roots, and you'll likely find Delbrück, Luria or Hershey. And, always, a phage.

Sadly, though, the term 'bacteriophage' remains unfamiliar to the general population. And as we'll see in the next chapter, our understanding of the true abundance, diversity and importance of these seemingly obscure viruses on planet Earth is only just beginning. While the work of the Phage Group has unlocked many secrets about how life works at the molecular level, we still know virtually nothing about the majority of phages on this planet.

# 13

## Planet Phage

In March 2012, the 36-metre schooner *Tara* arrived back to her home port of Lorient, in northwest France, flashes of orange livery on her bow and masts breaking up her unusual all-aluminium hull and masts. The brutalist laboratory-cum-sailing ship had survived expeditions round the iceberg-strewn seas of Antarctica and attacks by pirates on the Amazon River, but for the previous two years *Tara* had been carving tens of thousands of miles around the South Atlantic on a special microbial mission. The boat and its crew of researchers were making the first attempt to chart an entire ocean's microscopic life and its influence on the biosphere and its climate.

Within months, scientists sampling water from the trip had identified over 15,000 new viruses,[50] more than tripling the previous estimate of known viruses in the world. When the researchers analysed more water samples – this time with an improved technique for finding viral DNA – they identified almost 200,000 new viruses. Years after *Tara* arrived back in port, a team of international scientists is *still* analysing the thousands of water samples taken from the sea as the ship carved around the southern half of the globe. The more they look, the more viruses they find. The variety seems almost infinite – viruses of all shapes and sizes that infect every marine organism imaginable. But the vast, vast majority are phages.

As recently as the late 1980s, it was believed that the concentration of viruses in clean, open water was low and, therefore, the numbers of viruses in the world's lakes, seas and oceans was largely insignificant. In fact, virtually nothing was known about the abundance and diversity of viruses in the oceans, with the majority of virus research focused on those in our immediate surroundings that cause human or animal diseases.

That only started to change in the final few months of the 1980s, when a study published by a group of Norwegian researchers suggested microbiologists were missing something big when it came to understanding the role of viruses in the environment. Øivind Bergh and a group of colleagues based at Bergen University, on Norway's archipelagic western coast, decided to try and work out how abundant viruses were in large, open and unpolluted marine environments. While it was known that phages and other viruses could be found easily in dirty environments – like sewers or polluted rivers – few had bothered to look for viruses in clear and pristine waters. Those who had done, in the past, could only detect what could be grown in their labs. Given that fewer than 1% of aquatic bacterial species can be grown in the lab,[51] those represented just a tiny fraction of what we now know is out there.

Bergh and his group wanted to study waters untarnished by human activity and took samples from open water on both sides of the Atlantic – from the beautiful Chesapeake Bay on the East Coast of the US to Korsfjorden, a desolate fjord at the extreme northern tip of Scandinavia. They looked at tiny amounts of finely filtered water under electron microscopes and counted anything that looked like a virus. There were far more than they expected. They saw viruses everywhere, even in truly tiny amounts of water, many in the act of attacking and bursting bacteria.

The group's study, published in 1989 in the top scientific journal *Nature*, came to a remarkable conclusion: rather than the number of viruses in the oceans being somewhat insignificant, there appeared to be tens or even hundreds of millions of viruses in every millilitre of water in their samples, on either side of the Atlantic.[52] Given these samples were taken at random, from open water, did that mean there were hundreds of millions of viruses in every millilitre of water *in the entire ocean*? It would mean the amount of viruses in the sea and on the planet was preposterous. A few months later, in 1990, two more papers were published drawing the same conclusion – there were *billions* of viruses in a litre of water, this time scooped up way east of the Caribbean;[53] and similarly massive numbers were found in the hyper-salty Laguna Madre lagoon on the Texas coast and in the tropical open waters of the Gulf of Mexico.[54]

Marine bacteria were once believed to be the most abundant form of life in the oceans, but these papers were suggesting they were outnumbered by viruses ten to one, and in some environments one hundred to one. Further studies confirmed that viruses are, by far, the most abundant biological entities in the sea, 'nearshore and offshore, tropical to polar, sea surface to sea floor, and in sea ice and sediment poor water', according to one paper published in 1999. There are more phages in a typical litre of seawater than there are humans on the planet,[55] and there are over 1,000,000,000,000,000,000,000* litres of

---

* Or $10^{21}$ (1,000,000,000,000,000,000,000), another stupidly massive number known as a 'sextillion'. Given that some research suggests our brains can't really picture amounts larger than ten, don't feel bad for not really understanding what this means. It's many trillions.

water in the world's oceans. Rather suddenly, and surprisingly recently, we have come to realise that the number of viruses in the environment is not 'insignificant' at all: they are in fact easily the most abundant biological entity on Earth.

Over the last hundred years, as phages became central research tools in many Western bioscience labs, and the Soviets spent decades perfecting the mass production of phages for medicines – too few scientists looked up from their Petri dishes, brewing flasks or microscopes to consider a bigger question: how many phages are really out there, and what impact do they have on life on Earth as a whole? Because of our strange blind spot for phages, we still don't fully understand. Scientists who spent decades calculating how energy and nutrients are cycled through the ecosystems and environments of our planet never considered the contribution of this vast churning soup of viruses, estimated to hold roughly 200 megatonnes of carbon[56] at any one time. That's fifty times the amount of carbon emitted by human activities across the entire US each year. Despite their sub-microscopic individual size, there are so many phages on Earth that if they were all put end to end, they would stretch for 400,000 light years into deep space. The entire Milky Way is only 25,000 light years across.[57]

Marine ecologists studying food webs in the ocean had long believed most marine bacteria died by being eaten by slightly larger microscopic organisms called protozoa, or by accidentally being ingested by larger animals. They now had to reckon with the idea that up to half of the bacteria in the oceans were dying from viral infections.[58] When bacteria are eaten, much of the energy and nutrients within them enters the food chain, eventually being eaten by bigger and bigger organisms like whales and sharks. But for those that are infected and burst open, however, the opposite happens: the nutrients and energy spill out and dissolve into the sea.

The rules of ocean biogeochemistry have had to be completely rewritten. It is now estimated that this effect, known as viral shunt, releases up to three gigatonnes of carbon into the seas every year,[59] or ten billion tons per day.[60] The release of all this carbon and other nutrients into the water actually helps stimulate the growth of more bacteria, creating a vast energy and nutrient recycling system that has existed long before the evolution of larger and more complex life in the seas. One extremely rough guesstimate suggests the world's phages are popping bacterial cells open 100 billion trillion times a second. This liberates such huge quantities of bacterial innards into the sea that it forms a kind of marine snow, which is either rapidly taken up by other life or sinks to the bottom of the ocean.[61]

After the revelations at the start of the 1990s, the discoveries about viral abundance just kept coming. Virologists then began to explore the number of phages on dry land, for example in soil. Concentrations were *even higher*, with over a billion phages found in each gram of soil in wetland or forest environments.[62] Briny marshes, acidic lakes, caves cut off from the rest of the world for thousands of years: everywhere researchers looked, millions of phages. Even a teaspoon of earth from a baked desert, or ice locked into glaciers for thousands of years, can contain an active community of thousands or millions of phages.[63] When a group of researchers installed collection devices on a concrete platform almost 3km above sea level in the Sierra Nevada mountains, in Spain, they found hundreds of millions and sometimes billions of viruses were simply falling from the sky every day,[64] perhaps blown high into the air from sea spray or dust and transported around the world on global air currents. In 2022, scientists from the University of Copenhagen announced they had found 876 different types of phage *just on the leaves of wheat plants*, 848 of which were entirely new to science.[65] Other research from 2022 even suggests that phages may help certain unusual rock structures form from minerals and sediments in the sea, literally shaping the very Earth itself.[66]

Together, these studies have changed our understanding of our planet.* Scientists now believe that phages kill between 20% and 40% of any given population of bacteria in natural environments every day.[67] The same effects are seen in the mini ecosystems inside the intestines of all animals, where the many useful nutrients locked up inside their gut bacteria are continually liberated by phage infections to be absorbed and used by the animal or excreted into the ecosystem.

The well-established idea that microbes drive ecosystems on Earth has rather suddenly been reappraised: microbes may well rule the world, but viruses rule the microbes. The study of viruses in the environment, known as viral ecology, has gone from a curiosity to a crucial frontier in oceanography, climatology, biogeochemistry and marine biology. The more we look at these phages, the more stunning insights about their importance are revealed.

---

* Given the sheer abundance of viruses in the environment, it seems that something in nature should have evolved the ability to consume them. Just as this book was going to press, scientists at the University of Nebraska proved that a unicellular organism known as *Halteria* that can survive solely on a diet of viruses, the first example of 'virovory', or virus-eating, ever discovered.

Jennifer Brum was raised in Las Vegas, in the middle of Nevada's Mojave Desert. The relentless heat and family trips to the coast inspired dreams, from a young age, of being an oceanographer, a career where she could dip into the cool waters of the Pacific anytime she liked. Broad-shouldered, tanned, with a wide smile, Brum is now assistant professor in the Department of Oceanography and Coastal Sciences at Louisiana State University. She spends her time looking for viruses where nobody has looked before, from aboard research ships like *Tara* to local lakes and sewers.

As well as nineteen research cruises to different parts of the ocean, she has squelched through Louisiana's Atchafalaya basin, the largest swamp in the US; trekked up California's Sierra Nevada mountains to sample the hostile Mono Lake, 9,000ft above sea level, three times saltier than the sea and full of arsenic; and she has sent containers 2,500m below the ocean's surface to collect water from the mid-Atlantic ridge – where two great plates of the Earth's oceanic crust collide and superheated gases from the inferno below spew out from the mountainous wreckage.

Everywhere she goes, new viruses, new genes, new insights into how life on Earth really works. In the oceans' deep, dark low-oxygen zones – which worryingly are expanding both vertically and horizontally – Brum found phages which give their bacterial hosts genes that help them deal with the lack of oxygen. In Mono Lake, where the most complex life that can survive is a salt-tolerant shrimp, she found the highest concentration of viruses of any previously studied aquatic environment. Most of them are unlike anything else found on Earth. And at the bottom of the sea, she found viruses that she believes could help us understand the very earliest life on the planet.

Although there are many competing ideas about how and where life on Earth originated, the theory that it was near deep-sea hydrothermal vents is probably the strongest. The hot, mineral-rich chimneys of gases erupting from the vents combine with seawater to form strong alkaline and acidic fluids, driving the strange energy-hungry chemical reactions required to form complex organic compounds. Some of the oldest fossils ever found are tiny filaments made by bacteria growing around iron-rich hydrothermal vents almost four billion years ago. Brum believes the phages she collected from these bizarre underwater environments are a glimpse into the world's most ancient life forms.

'One of the theories that people have put forth is that viruses originated right along with the first cells,' she says. 'So, it's very interesting to me to study viruses in this sort of "proto-world" ecosystem. The first role that they had was probably not infecting or killing cells, it was just transferring genes back and forth. It's likely viruses had a very big role in structuring the evolution of life on our planet from the beginning – the very beginning. And some have even said that life on our planet wouldn't exist without them.'

Billions of years later, phages are still playing this crucial role today, constantly co-evolving with their bacterial hosts. Phages' relentless pursuit of host cells pressures bacteria to constantly change, making them the super-diverse and complex group of organisms they are today. What's more, phages often make copies of themselves that accidentally contain some of their host's DNA, instead of just their own, which then goes on to be injected into whatever cells its progeny go on to infect. On a planetary scale, and over billions of years, this transfer and mixing of genetic material has an enormous effect, constantly creating new and unusual combinations of genes. Viruses, rather than just parasites that kill their hosts, are increasingly seen as versatile carriers of genetic information within and between species, constantly rearranging existing genetic information into unique combinations.

As well as taking genes from their hosts by mistake, when phages insert their DNA into a host, they may also lend them genes to help boost their metabolism or fitness. Of course, the phage provides this genetic boost only to ensure that their host can produce as many phages as possible or reproduce as many times as possible, replicating the phage DNA many times with it. In some cases, phages pass on useful genes from one bacterial strain and give it to another, but in some cases, they have literally taken something the host already does and made it better.[68] Rather than bacteria having to invent their own solutions in the endless game of survival of the fittest, they can also borrow solutions from others – and phages are more often than not the intermediaries, helping spread useful innovations around the globe. This sideways transfer of genetic information, technically known as horizontal gene transfer, means the rules of evolution are very different for microorganisms: whereas us higher organisms have to wait to have sex to shuffle our genes and create variation in the next generation, bacteria and viruses and other completely unrelated lineages of microorganisms are able to swap, steal and share genes that others have spent millions of years developing. And thanks to phages, it is estimated that this genetic transfer between organisms takes place about 20 million billion times per second.[69]

The integration of phage DNA into bacterial cells can have vast and deadly consequences for us, turning once harmless bacteria into deadly killers. For example, the pathogens that cause cholera and diphtheria were once harmless marine bacteria that had little impact on the human body. But at some point in time, they were given genes by phages that turned them into deadly pathogens. In the case of cholera, strains of the bacteria *Vibrio cholerae* that have become chronically infected by the phage CTXφ are given the genes for the production of a particularly nasty toxic protein which causes the distinctive intense watery diarrhoea characteristic of cholera infection. Likewise, the bacterium *Corynebacterium diphtheriae* can only produce the potent toxin that causes the symptoms of dangerous childhood disease diphtheria after infection with a phage known as β-phage. Remarkably, outbreaks and epidemics of these diseases can be linked to epidemics of phage infections within the bacteria responsible.[70] Countless other strains of common bacterial species, from *E.coli* to *Staphylococcus aureus*, are given more subtle assistance in causing human infections by phages, including genes that help them evade our immune system or penetrate our tissues. Phages also help transport genes for antibiotic resistance from one bacteria to another.

The phages that carry these deadly genes, responsible for millions of deaths throughout history, make *The Good Virus* seem like a rather questionable name for a book about bacterial viruses. They also show how important it is to understand the biology of phages before we inject them into the body as medicines. If phages are to be used in medicine on large swathes of the population, regulators must be confident that they do not contain genes that could make the bacteria in our bodies even stronger and more deadly – hence why lots more research and genetic data is needed to complement the results of clinical trials as they progress. A slow and careful introduction of phage therapy into mainstream medicine – although agonising for those who need it now – will hopefully ensure that the viral medicines of the future are not harbouring dark and mysterious genes that make our multifaceted healthcare crises worse.

Mya Breitbart, professor of biological oceanography at the University of South Florida, is another of the world's great phage hunters. She began her career as a marine biologist with 'no interest in microbes beyond the minimum learning required on my marine science course'. Then she heard about a colleague who was developing new antibacterials from compounds found in Antarctic sea sponges, which the sponges use to ward off infections from marine bacteria, and learned of the crazy abundance of marine bacteria and their viruses.

'That was kind of what got me hooked,' she says, apologising for speaking so quickly as she's almost bubbling over with enthusiasm. 'I see so much potential for finding phages in environmental samples that can be useful to us, whether it's from a phage therapy standpoint or for new biotechnology enzymes. It baffles my mind how much is out there and how little we know about it.'

She has lost count of the number of carboys – heavy duty cylindrical containers for transporting liquid – she has filled to find viruses from various stretches of water around the world. For Breitbart and her fellow phage hunters, sampling for viruses is second nature when visiting a new location. 'I've done a lot of filtering in hotel bathrooms, bathtubs in particular because the water gets everywhere. I remember we went to a conference in Salt Lake City and all went out to the Great Salt Lake and pulled up bottles of water to sample, just, you know, because we were there . . . We kind of can do it when-ever, wherever we are.'

One of Breitbart's most read papers on phages is the intri-guingly titled 'Puppet Masters of the Marine Microbial Realm'. In it, she describes how phages dominate marine ecosystems, not just regulating different bacterial populations but signifi-cantly changing their hosts' metabolisms and behaviour. Phages can alter their hosts' metabolism and behaviour so much that some researchers think that bacteria of the same species that have and haven't been infected by phages might need to be considered different organisms.[71]

The most striking example of this is the ancient green-blue marine organisms known as cyanobacteria. They are probably the most abundant type of bacteria on the planet and evolved the ability to photosynthesise – using the sun's energy and atmospheric carbon dioxide to produce food and oxygen – long before plants did. We think of trees and forests as the lungs of the planet, but up to half of the oxygen in the Earth's atmosphere is produced by these bacteria, silently and mostly invisibly in our oceans. Cyanobacteria were central in the shift in Earth's atmosphere, around two billion years ago, from an anoxic hellhole to one rich in oxygen. We and all other animals would not be here without cyanobacteria. Being present in such large numbers, cyanobacteria are, of course, a target for many phages. The most abundant phage on the whole planet is likely to be one called HTVC010P,[72] which infects the most common type of cyanobacteria in the ocean, known as SAR11. (Yes, they both need snappier names.)

What is really neat about the phages of cyanobacteria is that when they infect their host, they give them extra genes that allow them to temporarily photosynthesise even more efficiently than they already do.[73] Bacteria infected with phages get a kind of turbo-boost, able to produce more energy and oxygen from less light. From the phages' point of view, this is purely out of self-interest – this genetic boost ensures their hosts have more energy for making new phages. With up to half of all bacteria in the oceans infected with phages at any one time,[74] this can have an enormous cumulative effect on the planet. Researchers now believe as much as an eighth of all the oxygen we breathe was produced thanks to this phage-based boost to marine bacteria's metabolisms. Scientists like Breitbart are only just starting to study how all these viruses and their interactions with their hosts may change as our climate warms.

In the brutally cold seas of the Antarctic Ocean, the formation of new sea ice in winter sets up the physical conditions for enormous blooms of photosynthetic bacteria, known as phytoplankton, to form when the ice recedes in summer. The seas become thick with these green, light-harvesting microorganisms and the growing bloom draws $CO_2$ out of the atmosphere and into the thousands of different species of animal in the food chain above it.

In these freezing polar regions, many phages live quietly inside their bacterial hosts for most of the year, only switching to the more frenzied 'reproduce and kill' mode when the blooms begin, when they can run rampant through a densely packed population. This killing frenzy bursts vast numbers of phytoplankton open and liberates huge quantities of carbon and nutrients from the cells into the water and local ecosystem. But, of course, the amount of sea ice that forms each year is now declining each year as the world warms – and the scale of these important phytoplankton blooms are falling too. Scientists are now asking: if the bloom never comes, do the phages never switch into kill mode? Does less sea ice mean smaller blooms, lower phage infection and less carbon released into the water? The interplay between climate change and phages in just this one region remains unclear. But what is clear is that phages – an enormous churning and largely unaccounted for mass of microscopic predators – could have a wide range of potential effects as the climate changes; either driving the liberation of more carbon from marine stores and driving further warming, or helping the oceans absorb more. And while these questions will occupy scientists for many lifetimes, other scientists are already looking beyond our planet – at the phages we have unintentionally put into outer space.

In NASA's famed Kennedy Space Center, on the vast, flat grasslands of Cape Canaveral, spacecraft spend their final days on Earth being assembled and inspected in giant high-security spaces known as clean rooms. Engineers work in full head-to-toe protective white suits, with goggles, plastic boots and gloves over the top, only entering the brightly lit and spotless hangars having undergone rigorous decontamination procedures on the way in. The outward facing surfaces of the spacecraft are baked at temperatures of up to 350°C on their way in to ensure any microbes on them are vaporised.

This is not just about making sure that NASA's delicate and expensive technology is free from any microscopic pests that could affect its operation or protecting the health of the crew. This is about protecting the rest of the universe from invasion by Earthly life forms.

NASA takes the idea of contaminating other worlds with

our microbes pretty seriously. They have a dedicated Office of Planetary Protection, which promotes the responsible exploration of space and helps develop their processes for 'protecting solar system bodies from contamination by Earth life' – as well as protecting us from extra-terrestrial life forms that may come back to the planet.

Gary Trubl is a square-jawed and athletic-looking Arizonan who by day studies phages in extreme environments, but by night increasingly worries about the phages we have unintentionally sent to other worlds. After years spent studying the super hardy bacteria and viruses locked in oxygen-free permafrost, or in the sludgy brine stuck beneath 13 metres of Arctic ice, he now wonders whether these 'extremophile' microbes could also survive a trip into space. Studies of supposedly sterile NASA clean rooms have revealed phages often escape NASA's strict clean room decontamination procedures,[75] and Trubl believes phages have almost certainly hitched a ride to the moon, and Mars, and may even be zooming at tens of thousands of kilometres an hour into deep space as we speak. (Voyager 1, the farthest human-launched object from Earth, is currently almost 23 billion kilometres away at the time of writing.) One famously strange and speculative article (famous in the phage community, at least) once even suggested that phage φX174, one of the simplest replicating life forms known to science, could be an interstellar message sent from an advanced extraterrestrial civilisation.[76]

'The biggest question in all of humanity is "are we alone?"' says Trubl, who talks slowly and earnestly through pearly white teeth. 'If we go somewhere looking for life, we don't want to litter the place with our own life – then we'll never really be able to answer our question. Then there's the fact that if whatever we take there has survived, we're introducing an invasive species into a new ecosystem. And when we do that on Earth, it's usually been disastrous. And we know that we're not doing such a good job – we've routinely found viruses in spacecraft clean rooms.'

Trubl's efforts to get NASA to pay more attention to viruses in space are not just about the somewhat remote possibility that we find life on other worlds: it also has direct implications for the health of astronauts and civilians who now go to space. Astronauts returning to Earth after a period in space have experienced worrying 'reawakenings' of previously latent or dormant viruses in their body, including strange rashes, flus and herpes. It's not clear if this is related to the conditions in space, such as zero gravity, or the stress of space flight, or both – but it needs to be understood as a matter of urgency if humans are going to embark on long- or medium-term space exploration missions.

'If you're up in space and halfway into your mission, and all the latent viruses in your body start to emerge, that's a real problem. And if you're taking plants or algae up in space to produce food or oxygen, and a virus within them suddenly reactivates and wipes them out, that's a disaster.'

The more Trubl has looked into the topic, the more he found that the study of phages and other non-human viruses in space was being neglected, just like it has been on Earth. Not just by NASA, but by the whole field of science that studies life beyond Earth, known as 'astrobiology'. He says there has been a reluctance to study viruses in space or look for them on other worlds because they are not considered alive, have traditionally been hard to find and study or because the public aren't interested in them.

'We need to have the machines that look for life beyond Earth set to look for viruses,' says Trubl. 'Every living organism we've ever seen, we've found that it has a virus, so far. We haven't found life without viruses, and we have never found a virus that can survive without a host. So, to me finding a virus is indicative of a host as well. What's also interesting is that viruses are our planet's most abundant biological entity. So, if there was life somewhere else, we would be more likely to find them than anything else.'

Searching for bacteriophages in space, at the bottom of the sea or in acidic hypersaline lakes might seem to some a little abstract – irrelevant, perhaps. But this work is in fact already generating brilliant insights on how phages interact with the bacteria in our bodies and how that impacts our daily lives.

Forest Rohwer looks more like a former Grateful Dead roadie than one of the world's leading viral ecologists. With long grey hair and messy grey stubble, often kitted out in black T-shirt and black boots, he speaks in a surprisingly soft and slightly squeaky West Coast tone. Decades ago, Rohwer translated his love of nature and hunting – from big game tracking to spearfishing – into a career spent studying marine ecosystems – coral reefs particularly – and hunting viruses. When I ask him where he's been to study phages, he answers with a smile and an eye-roll: 'Like, the whole world.'

Despite his surfer dude demeanour, Rohwer has been instrumental in developing the complex molecular techniques that now allow virologists to 'find' tens or even hundreds of thousands of different viruses at a time from samples of water, without the need to ever grow or see them. Known as 'shotgun metagenomics', the process involves chemically isolating all the viral DNA in a given sample and analysing it together, using computer analysis to estimate which of the DNA sequences in there represent unique viral species.

It was these genetic techniques that caused the number of known viruses in the world to spiral wildly from the early 2000s, from a few thousand to tens of thousands, and then with expeditions like *Tara*, to hundreds of thousands. As these results poured in, each one revealing more and more completely new viruses, it led some scientists to predict that there may be trillions of different species* of phage in the world.

I ask Rohwer if he thinks it is even possible for researchers like him ever to get a handle on the full diversity of phages out there. In 2005, he published a paper called 'Here a Virus, There a Virus, Everywhere the Same Virus', in which he argued that perhaps all the different phages in the world have been so well mixed by ocean currents that the diversity isn't quite as mind-boggling as we think. So while you might find hundreds of thousands of different types of virus in a single litre on one side of the Atlantic, a litre on the other side of the world might contain *roughly* the same hundred thousand viruses; what matters in any given location is which of those phages are actually active, and which are present in large

---

* Phages that are 95% genetically identical (or more) are considered by scientists to be the same 'species'. But that figure is somewhat arbitrary – just something that allows scientists to group highly similar phages together and give them a name. Many virologists argue that the concept of species doesn't apply to viruses anyway – their jumbled-up and ever-changing genomes just don't relate to each other in the same way as the genomes of more stable organisms that live longer lives and replicate themselves more cautiously.

numbers. If this is the case, researchers should at some point be able to get a handle on all the different types of virus there are on the planet – although that may still be some years from now.

But that was 2005. Rohwer is now working on a theory that maybe there is not a quantifiable amount of phage species out there for scientists like him to catalogue and put a number on. Instead, he is starting to think the immense diversity found so far may be a result of new phages constantly 'coming into being' in a way that we don't yet understand. Does he mean evolving into new species, but very quickly? I ask. No, he says, with a pause for added mystery. 'Entirely new viruses may be emerging from the microbial ether more suddenly even than that', he says. What could be driving that? I respond. 'Probably magic,' he says, laughing tiredly at the enormity of the question. 'When you're hunting for something new like that you can't know what you're looking for. You just have to design the way of hunting that helps you find it.'

If you're beginning to think Rohwer's work sounds esoteric and obscure, it isn't. In fact, it is one of the rare fields where our understanding of marine systems is ahead of our understanding of human systems, rather than the other way round. Rohwer's work has had a direct impact on helping us understand how the natural ecology of bacteria and viruses can affect human health.

Take his work on the phages living among the beautiful corals of the Pacific and Indian oceans for example. His years studying reefs around the Caribbean and Sri Lanka have showed that bacterial colonisation of reefs is a primary factor in their decline around the world, and that environmental stresses caused by humans, such as overfishing, is likely the cause of these underwater bacterial infections. The bacteria that live on the coral are themselves infected by a variety of specially adapted phages – and so phages could have a major role to play in saving coral reefs around the world, or at least slowing their decline, by keeping harmful bacterial colonisation in check.

But here's where it gets really interesting. In the early 2010s, Rohwer and his then-postdoctoral student Jeremy Barr observed that there were extremely high numbers of phages in the slimy mucous layer found on the surface of corals – up to four times as many as in the surrounding environment. These mucous surfaces form a kind of sticky protective physical barrier between the interior of the coral and the environment, but also provide ideal conditions for bacteria to stick to, feed on and colonise. They are surprisingly similar to the mucous surfaces found in the lungs and guts of many animals, including humans.

Rohwer and Barr discovered that phages living near these reefs had evolved the ability to adhere to and penetrate the corals' mucous layer. The reward for the phages that could penetrate the mucous was access to lots of bacterial hosts immobilised in the mucous, and therefore greater reproductive success. Artificially removing the phages increased the chances of the coral beneath the mucous layer becoming infected with bacteria. So, in return for access to lots of potential hosts, the phages were helping reduce bacterial growth on the outer surface for the coral.

The researchers quickly found this neat symbiotic relationship was happening not only in corals but also in the mucous of more complex marine organisms like sea anemones and in fish. In all these organisms, phages were acting like live-in security guards, packed into their mucous in high numbers and helping stop bacteria from colonising these important barriers between their insides and the outside worlds. As the organisms co-evolved, the marine animals had made their mucous more attractive to the phages. In return, the phages were providing them with a kind of external immune system.

Rohwer's student, Barr, now head of the bacteriophage research unit at the University of Monash in Melbourne, began to wonder if the same phenomenon was occurring in the human body too. And, perhaps unsurprisingly at this point, it was: the same molecules that helped the phages stick to mucous in the sea could be found in the mucous surfaces throughout the human body.

Back in the 1920s, one of Felix d'Herelle's more outrageous claims was that phages were an integral part of our immune system, a kind of live-in virus that our bodies had co-opted to counter bacterial infections on our behalf. He was widely derided for the idea and, until just a few years ago, it was still widely believed that phages did not interact with animal cells. But now, almost a century later, Barr's lab in the suburbs of Melbourne is focusing specifically on what he now calls our 'third immune system': the ecosystem of phages that protects us from bacteria alongside our other, more well-studied innate and adaptive immune systems. Barr's work has upended the idea that phages do not interact with mammalian cells, telling *Science* in 2017 that the notion was 'BS'. He has even observed phages being engulfed by human cells and trafficked around the body. The exact way this is achieved is unclear, but the work tantalisingly suggests that our bodies actively co-opt and deploy phages where they are needed like working parts of the immune system. His work also suggests that through our guts, we absorb as well as host tens of billions of phages each day. Phages seem able to freely penetrate our bodies; ending up in areas of the body once classified as 'sterile' and lacking microbes, for example the blood, lungs, liver, kidney and even within the brain. It goes a long way to explaining why phages only sometimes illicit an immune response when administered therapeutically – our bodies have co-evolved with these viruses and rarely see them as foreign objects, unlike other viruses.

So, we know now that the delicate barriers that separate our bodies from the outside world appear to have evolved to be especially welcoming for phages, and that our bodies interact with phages in many ways once thought impossible – all thanks to phage ecologists studying seemingly obscure marine viruses at the bottom of the sea. Barr is now studying how phages modulate the delicate balance of bacteria in our intestines, known to be associated with many different types of poor health, in the hope that one day phages can be used to reverse or correct disorders of the gut.

Meanwhile, Rohwer is now using his understanding of the complex microbial diseases of coral to tackle cystic fibrosis, a disease in which the primary problem is an excess of mucous, which then leads to extremely persistent infections. 'Since we're already working with a lot of mucus, it makes sense,' says Rohwer nonchalantly. His knowledge of infections on coral reefs can be applied to the complex community of viruses and bacteria in the lungs of cystic fibrosis patients, and their dynamics, helping us see the lungs of cystic fibrosis patients as entire ecosystems, not just as unhealthy organs.

Further studies of the ecology of phages in complex ecosystems – captured in the messy real world, rather than in carefully controlled lab experiments – is providing important insights for those hoping to use viruses to treat infections. The warm, wet environs found in and on the human body are complex microbial ecosystems too, after all.

In 'the wild', phages rarely completely eradicate populations of their hosts: to do so would mean they have no cells in which to replicate. Instead, microbial communities exist in a state of permanent flux, where dominant bacterial strains become subject to intense viral attack and less numerous strains are left in relative peace (or with phages lurking quietly inside them). When the dominant strain is killed to the point that its abundance declines below a certain point, the amount of phages available to kill them declines too. Another bacterial strain starts to dominate and the phages of that particular strain start to proliferate. This dynamic, known as 'kill the winner strategy', tells us that a single type of phage is unlikely to ever completely eradicate a bacterial pathogen from our bodies – but it could degrade the population to the point where our immune system, or antibiotics, or different phages, or a combination of all three, finish off the remaining bacteria and clear the infection. If all this has been gleaned from just over thirty years' worth of work on phage ecology in a few marine and soil ecosystems, imagine what we could learn from the rest of the phages out there in the world.

For now, a surprisingly small but passionate community of viral ecologists are tasked with a near-impossible job: finding, analysing and classifying the most abundant and diverse life form on the planet. The vastness of their task is hard to comprehend: this super diverse and hyper-abundant group of viruses represents the largest reservoir of unexplored genetic information on the planet. With potentially trillions of different types of phages out there, many refer to them as the 'dark matter of biology'.

Understanding the true number of viruses in any given place, what they are, and what they are doing remains an exceptionally difficult task. It is not just the sheer numbers that make it difficult – phages have several unique characteristics that can make them particularly elusive even to our most advanced technology. The most basic type of phage hunting, simply seeing what forms plaques on dishes of different lab bacteria, will only ever identify phages which infect the bacteria that can be grown in a lab. The majority of bacteria – some microbiologists believe as much as 99% – cannot be grown in lab conditions, and so their phages will never be found this way. Even in those phages that do infect the bacteria in your lab, not all will form plaques, and so they too will remain unseen and unstudied.

If you decide to painstakingly count everything that looks like a virus in your sample, either by eye or with sophisticated laser detectors, you will miss the many viruses that are temporarily or permanently residing inside a host cell, known as prophages. And as mentioned earlier, you might miss those that are much smaller or larger than our expectations of what size phages are. If you search for viral DNA sequences, you miss all the viruses that use DNA's chemical cousin, RNA, as their genetic material.* Then there's the problem of phage genomes falling apart in the time between sampling and analysis.

What further complicates the search for phages – other than the fact that there are trillions of them, and they are

---

* Databases of known phages are dominated by DNA-based phages rather than RNA-based ones, and just two of the hundred or so major phage 'families' have RNA genomes. But because over many years, most researchers have been looking for DNA phages, the true proportion and diversity of RNA-based phages is unclear.

always changing – is that they are what is known as *polyphyletic*. This means that there is no clear ancestral lineage from which all modern phages evolved. Phages appear to have emerged and evolved completely independently from one another at many different points in the history of life on Earth. This lack of a common root means that there is no one universal gene or genetic sequence that all bacteriophages have – nothing that allows scientists to easily say 'this DNA comes from a bacteriophage'. As a result, there are almost certainly phages out there with DNA sequences so unlike anything ever seen before that even computers programmed to identify phage-like DNA sequences wouldn't recognise them. It may even be that the *majority* of phages fall into this category – too different from any known virus to be recognised by our best technology as a virus.

As a result of these challenges, many believe our best research is still only picking up a fraction of the true diversity present in the samples, and that there could be as many as ten trillion different species out there waiting to be found. Approximately 70% of the total volume of the oceans is located deeper than 1,000m, an area that has been sampled far less frequently than the upper ocean. Then there is a whole other group of single-celled organisms, similar to bacteria, known as archaea. This extremely diverse yet understudied group of ancient microbes also have viruses, about which even less is known. These viruses, also known as phages and grouped with bacterial viruses simply for convenience, are *even stranger*, forming unusual shapes and structures not seen anywhere else in nature. In one bizarre case, one deep-sea phage makes its archaeal host cell swell into a giant virus factory 8,000 times its normal volume.

It's often hard to imagine what we might learn, or how we might benefit, from the painstaking study of utterly obscure organisms in obscure environments. Many will wonder if examining endless viruses found in weird places is worth the effort. But there is one powerful example of why curiosity-driven research is so important; why research with no other goal than understanding, and where the 'real world' uses are not immediately apparent, is something we have to keep doing. It's a story about how a few people asking an obscure question about an obscure microbe snowballed into arguably the most powerful technological breakthrough of the twenty-first century.

# 14

## An ancient technology

The town of Santa Pola lies a few miles south of Alicante, on Spain's southeastern coast. Here, the ocean lies trapped behind a strip of beaches and roads connecting the local headlands, with shallow stretches of salty water and marshland extending for miles inland. Under the hot, often cloudless skies, a vast grid of paths separate shallow lakes used for salt production. Flamingos pad across the horizon, wobbling in the hazy heat, while enormous piles of salt look like icebergs, rising impossibly out of the arid land.

In 1989, a twenty-eight-year-old graduate student, Francisco Mojica, born just a few miles away and studying at the nearby University of Alicante, was trying to find out more about how life survives in the hypersaline waters of Santa Pola. His particular interest was an obscure microorganism that had barely registered in the public consciousness at all. Not phages. Not yet.

Mojica is now a famous name in the phage world, but all those years ago, he was looking for microbes called archaea. This diverse group of organisms are, to the untrained eye, just like bacteria: tiny single-celled microorganisms that replicate asexually. But they are, in fact, an entirely distinct lineage of microbe, with fairly fundamental differences in their basic biochemical building blocks. Many scientists believe that all of Earth's multicellular life – fungi, plants, animals, us – probably evolved from these ancient and less well-known microbes, while bacteria took a separate path, evolving in their own way on the opposite side of the evolutionary tree of life.

The species of archaea Mojica was particularly interested in was *Haloferax mediterranei*, a microscopic blob-shaped archaea found only in extremely salty environments like Santa Pola. The species thrives here, where the water contains enough salt to kill most microbes within minutes. Unbeknown to Mojica at the time, the archaea that he was studying spent their lives bathing not just in salt, but in phages.

Phages that infect archaea are similar to those that infect bacteria, and both likely evolved from viruses that infected the ancient ancestral microbes that were among the first life on Earth. But archaeal viruses are perhaps even more unique; as well as circular or tailed, they can be bottle-shaped, spindle-shaped and even two-tailed. But when Mojica was studying archaea in the late 1980s and early 1990s, nobody knew or cared much about the viruses that might attack them. 'We didn't know what we know now,' Mojica tells me from his modest office in Alicante. 'That these environments have some of the highest concentrations of virus particles and virus-like particles.' Gently rubbing where his hair used to be, he tries to remember how many gazillions of phages there are in the waters here. 'Between 10 and 100 viruses per archaea, I think . . . so 10 to the nine virus-like particles per millilitre of water? I'm very bad with these numbers.'

I'm secretly glad someone else struggles with these extraordinary quantities. That's a billion virions* for every 1ml drop

---

* Or as researchers prefer to call them in this context, 'virus-like particles' given that most of them are as yet unstudied and unknown.

of water in these vast salt lakes. A billion viruses in every milli-litre of water that even microbiologists studying these waters didn't know were there.

'Viruses, in my field at the time, were just not relevant,' continues Mojica. 'Only a handful of archaeal viruses were known.'

In environments all over the planet, viruses massively outnumber their microbial hosts. It could be said that for most of the Earth's history, life has simmered in a stock of viruses, and likely has done so since chemistry first became biology around four billion years ago. This sounds like bad news if you are a bacteria or archaea, growing up in a world of infectious agents hoping to replicate inside you and then burst you. But the sheer number of bacteria and archaea happily replicating in any environment at any one time clearly suggests they know what they are doing when it comes to repelling these viral attacks. Without any kind of defences, single-celled life on Earth might have been wiped out long ago as their tiny foes ceaselessly bloomed at their expense.

Back in 1989, Mojica cared not a jot about the viruses surrounding the organisms he studied. He was tasked with trying to understand how his archaea were genetically adapted to surviving in the hypersaline waters of Santa Pola. Organisms that live in extreme environments are particularly interesting to scientists because they can provide clues as to how life survived in very early ecosystems on Earth and how life might work on alien worlds.

Mojica took samples of the microbes from the scorching salt lakes, broke them apart and purified just the DNA. Along the length of a DNA molecule, four different chemical subunits – adenine, guanine, cytosine and thymine, represented by the letters A, G, C and T – are repeated many, many times. It is the exact sequence of these repeated subunits – sometimes millions or billions letters long – that form every organism's unique genetic code or genome. And so Mojica set about working out the sequence of As, Gs, Cs and Ts that were unique to *Haloferax mediterranei*.

Mojica wanted to know if these DNA sequences would reveal anything interesting about how microbes live in such extreme environments – whether *Haloferax mediterranei* had genes that matched other salt-loving organisms, for example. Despite the tremendous advances in molecular biology set in motion by the work of the Phage Group and its protégés, in the late 1980s DNA sequencing was still horrendously laborious and time-consuming. After the long and delicate chemical procedure to break and pull the DNA apart so its constituent parts formed dark bands on X-ray film, Mojica had then had to translate thousands of these bands, each one representing an A, G, C or T, into a string of letters on paper, by hand. All this to understand the very long, potentially meaningless DNA sequence of a very obscure type of microbe. As the biotechnology pioneer Eric Lander would later write, Mojica 'could hardly have chosen a topic that was more local or less sexy'. But to Mojica, it was thrilling: decoding nature's blueprint for life in an extreme environment, letter by letter.

One day, amid the eye-straining jumble of letters, Mojica noticed something curious: the same thirty-letter DNA sequence seemed to pop up time and time again, repeated at regular intervals. After initially thinking this was some kind of mistake in his process, more sequencing yielded the same results: the DNA of his salt-loving microbes contained fourteen repeated sections of the same thirty-letter sequence, including palindromic bits, or DNA sequences that read the same backwards as forwards. The sequences did not match any known microbial genes, although one research group in the 1980s had seen a similar structure of repeating palindromic sequences in the bacteria *E.coli*.

Between the repeated sections were highly random sequences of letters, which also didn't seem to relate to any known genes. Mojica called them 'spacers'. The name unwittingly suggested this DNA was mere genetic filler between the interesting bits. They were to turn out to be anything but.

Mojica was baffled but intrigued, and his early ideas on what function these weird sequences of DNA might serve were speculative and wrong – at first, he hoped it might be something to do with how they survived in salty water.[77] He applied for funding to study them in more detail. Not only was he rejected, but the reviewers of his application actually complained about the sheer obscurity of his work, thereby damaging his reputation and ability to fund his work for years to come.[78] Finally, the Spanish government agreed to fund a study of the strange repeating sections, but only if he moved onto a well-known model organism like *E.coli*.

In the 1990s, as DNA sequencing technology improved, the number of DNA sequences being read and published by scientists around the world began to grow. Research groups began to add the sequences from the various species they worked with into large open databases, which Mojica checked frequently to see if other microorganisms had the strange repetitive sequences. He eventually found that they were present in over twenty different microorganisms, including medically important bacteria like *Mycobacterium tuberculosis*, *Clostridium difficile* and one of d'Herelle's old favourites, the bacteria that causes plague, *Yersinia pestis*. The fact that the repetitive sequences were present in both archaea and bacteria – the two great lineages of microorganisms that separated billions of years ago – suggested this was something fundamentally important to microbial life. But Mojica still had no idea what it did.

With a handful of research groups around the world now also interested in the same question, and people using different terms, Mojica wanted a name they could all use. Describing the characteristics of his sequences literally, he went with 'Clustered Regularly Interspaced Short Palindromic Repeats'. It was an ugly name, the kind you find only in the most technical papers on microbial genetics. But it formed a neat acronym that millions of people have now come to know very well, even if they don't have the foggiest idea what it actually means: CRISPR.

Just within the confines of a laboratory flask, within a few days or even hours, a bacterial cell can emerge with random mutations that help it to thrive in the presence of a virus that kills the cells around it. A few days more, and strains of virus will emerge that are more suited to taking that newly resistant strain down. It benefits both parties to change rapidly, ensuring a greater chance of evading the other's latest move.

This rapid and endless cycle of infection, mutation, selection and adaptation plays out trillions of times every second on Earth, and has done since the Earth was a fiery, noxious hellhole. The endless arms race between phages and their hosts has driven and continues to drive the diversity of both virus and host. From endless permutations of mutations in both predator and prey, pitted against each other each second of each day, in the survivors a beautiful intricacy and even a kind of intelligence emerges. Over time, the virus and its microbial host have both evolved remarkably complex and innovative chemicals in a kind of molecular warfare against each other.

As we have seen, bacteria have evolved an extraordinary range of defences against phages. As a first line of defence, they can prevent phages from attaching to them by changing the molecules on their outer coating that phages target. They may also block the phage's attachment site with decoy molecules or by secreting thick mucus-like substances to physically prevent the phage reaching their perimeter. Some bacteria even suddenly ditch the pulsating hairs (known as pili) or swishing tails (flagella) that they use to move, because phages use them to grab onto, before sliding down and infecting their body. Other bacteria allow infection with one relatively mild type of phage to prevent infection by another, more violently exploitative phage. In more extreme measures, bacterial cells will simply commit suicide once they detect infection, a selfless process known as 'abortive infection' where the cell effectively poisons itself, taking the phage genes down with it. In sacrificing itself to avoid being turned into a phage-factory, the bacteria helps save neighbouring cells from the same fate.

Phages, of course, are wise to many of these tricks. While the hosts modify the receptors to which the phages bind, so too the phage can also modify its own binding site, changing to recognise newly altered receptors or switching to target different receptor molecules altogether. They can also degrade or penetrate the mucous substance extruded by bacteria to keep them out. Remarkably, a certain strain of T4 phage found in the quaint waters of the River Cam, in Cambridgeshire, England, was found to be able to neutralise the toxin bacterial cells used to poison themselves, effectively preventing their suicidal hosts from killing themselves. Who knew such fraught microbial dramas could be playing out below the groups of tourists punting downriver?

There are even bacterial defence systems that guard their defence systems from counterattack by phages – a missile sent up to defend the anti-missile defence system from a missile. Once a virus breaches a bacterium's cell wall, the host still has options to prevent being taken over by the virus. So-called 'restriction enzymes' buzz around the cell looking for foreign DNA and chopping it up. These enzymes come in thousands of different varieties, and since their discovery in the 1970s have been used to cut DNA into very specific fragments in order to help study and manipulate genes. They quickly became the bedrock of most molecular biology and genetic engineering technologies, in fact. But as the turn of the century approached few biologists would have guessed that all that time, another, even more sophisticated bit of genetic biotechnology was lurking inside bacteria and archaea, just waiting to be discovered.

By the early 2000s, there were still only a handful of researchers who knew what CRISPR was, let alone studied it. Mojica's initial ideas about what these DNA repeats might be for had proven to be wrong. Still toiling away with huge reams of As,

Gs, Cs and Ts written across even huger reams of paper, Mojica had decided to turn his attention to the segments of DNA between the repeats: what he had originally dismissed as the 'spacers'. Working long hours in his Alicante office, he would occasionally open BLAST, a large online database of DNA sequences shared by researchers from all over the world, and type in the seemingly random jumble of letters he had found in the spacers, optimistically hoping for a match. There were none. But the database was constantly being populated with new sequences from researchers around the world, so every now and then he would search again. He had in fact been doing this for a decade, through the mid-1990s and into the new millennium.

'Ten years is a lot of time trying to find something,' he says. 'We knew sequences were being added all the time to the database that might match. But it really was just—' he searches for the word in English.

'Desperation?' I ask.

'Desperation, yes.'

To his amazement, in 2003, there was an exact match. The DNA sequence in one of the spacers he found in the genome of *E.coli* was identical to a sequence found by another researcher in a phage. Known as P1, the phage was known to attack *E.coli*. Intriguingly, Mojica's strain of *E.coli* was completely resistant to P1. Could the presence of P1 DNA in the *E.coli*'s genome be related to its ability to resist it? Did these bits of phage DNA, nestled amid the weird repeated segments, allow the *E.coli* to 'remember' the phages it had come across before? If so, that would be something massive. That would suggest that an organism's DNA is not just a blueprint for its life and functions, but also that it can form a kind of molecular memory; a storage device to remember the DNA of past attackers and prevent them from coming back.

Working manically, Mojica slogged through the DNA sequences in over 4,500 more spacers found in various other bacteria and archaea.[79] Many more contained fragments of DNA that matched sections of DNA from phages. He was increasingly sure of it – the CRISPR regions of bacterial and archaeal genomes were forming a kind of immune system, which somehow protected the microbes against phages they had encountered before. He allowed himself a celebratory night on the cognac with colleagues, before drafting an article on his exciting discovery the next day.

Frustratingly, his paper was rejected by four prestigious journals, one after the other, including *Nature*. Some journals did not even send his paper out to be peer-reviewed, known as a 'desk-reject' – the academic equivalent of a door being slammed in your face.

Mojica was sure he was onto something, but he did not have the experimental proof that the similarities between the spacer DNA and phage DNA had any particular function. It was a frustrating time for the modest man from Alicante.

For scientists in this position, there is the constant fear of being 'scooped' – in other words, another research group publishing the same idea or results and effectively getting all the credit for finding it first. Mojica began to despair. Other groups, including researchers at the French Ministry of Defence, were on the cusp of publishing similar findings.

'As you can probably hear, my English is not so good,' he tells me in perfect English, albeit with a pleasantly scrapey Spanish H. 'I was quite convinced this was something great. But I was not able to sell the idea properly. My former boss told me after reading the paper we eventually published, "you had gold in your hands but you weren't able to tell people".'

Finally, in February 2005, Mojica's first article on CRISPR's function was published by the *Journal of Molecular Evolution* – a rather specialist and not well-read journal. The competing groups published their results just a few months later, having also battled against rejections from the major journals.

Even after publication, CRISPR remained completely under the radar of most biologists. It's hard to overstate how unimaginably obscure and unloved this topic was in the mid-2000s: this was a poorly understood, highly technical question about the genomes of bacteria and archaea and phages, known to only a few groups of researchers – and of course there was that ugly and difficult-to-say name. It needed someone to elevate the field into the rock star topic that it would become. Enter an intense French researcher with 'CRISPR' licence plates on his Honda Accord and a wild-eyed energy for connecting his three scientific passions: bacteria, bacteriophage and fermenting food.

In the late 1990s, while Mojica was painstakingly matching CRISPR sequences to segments of phage DNA, a punky-looking, spiky-haired Frenchman called Rodolphe Barrangou was working with pickles. Industrial pickle manufacturers need microbiologists like Barrangou to monitor and maintain the fermentation of their vegetables – and it is not an exact science. 'We're talking, like, football fields' worth of vegetables in wooden tanks, with rainwater coming in, bird excrement coming in,' Barrangou tells me fondly. Analysing and regulating this pungent mix – rich in all kinds of organic material and microbes – to create a consistent product was 'like a mix of art and science,' he says. 'That's why I loved pickles so much. It was like making wine in the 1700s.'

Despite his passion for pickling, Barrangou's interests eventually turned to more precision food technology. As he developed skills in the rapidly evolving techniques of DNA sequencing and genome analysis, he was soon hired by Danisco, a brand of the chemical giant DuPont. They wanted him to find out more about the genes of their most commercially important bacterial strains, like the starter cultures the company supplies to yoghurt manufacturers all around the world.

Making yoghurt is not at all like making pickles. Sterile milk is inoculated with a precious strain of bacteria that has been selectively bred over hundreds and sometimes thousands of years to produce a perfect fermented product. Instead of leaky wooden tanks in a field, industrial yoghurt production is all vast stainless-steel vats and the kind of decontamination procedures you see in labs working with dangerous pathogens. Phages are the invisible enemy of this industry, where a single viral particle can wreak havoc, infecting and spreading through the all-important bacteria as it blooms in a batch. Any change in the speed of fermentation will affect the taste, texture and safety of the final product. Therefore, companies like Danisco invest enormous sums of money trying to find ways to defend their products from phages.

Barrangou, studying the genes of DuPont's highly valuable starter cultures, was especially interested in why some strains were resistant to phages and others weren't. For Danisco, a bacterial starter culture that was reliably resistant to phage attacks would be the holy grail of yoghurt production and potentially worth billions of dollars. Crucially, Barrangou and Danisco had something that researchers like Mojica didn't: a meticulous frozen archive of all their commercial strains *and* the phages that had caused them problems, going back decades.

'There is a very famous starter culture called 7710,' says Barrangou, who has a kind of hyperactive gum-chewing cool that Brad Pitt would play well if the story of CRISPR is ever dramatised. 'Someone would call us and say "hey we have a problem with it" and so we'd visit, get a sample and isolate the phage responsible for the trouble. Then you go back to the lab, expose the strain to the phage and see what grows. 7778 was made that way in '91.'

Thanks to this meticulous record, Barrangou had three crucial key pieces of the puzzle: strains of bacteria before they had a troublesome encounter with a phage; strains of bacteria that were somehow still able to grow after; and a sample of the phage responsible. Sequencing the DNA of all these elements of the puzzle, he suddenly had conclusive proof: the bacteria that were resistant now had bits of the phage's DNA incorporated into their own DNA, which had not been there before. Crucially, as double proof, bacteria that had been exposed to the phage, but did not develop any kind of resistance, did not. It was becoming clear that the CRISPR regions of archaeal and bacterial genomes were not genes but a kind of chemical library, where DNA from viruses encountered previously was deposited and held safely between the strange repetitive DNA sequences Mojica noticed on his reams of paper in the late 1980s.

For Barrangou, it was a breakthrough that would revolutionise the world – of yoghurt. 'Danisco Breakthrough Could

Boost Cultures' Resistance' was how his research was reported by the specialist food industry press in 2007. Again, it barely made a ripple anywhere else. As the number of researchers working on CRISPR grew, and their work became more confident and exciting, still they were met with resistance and rejections from journals.

'Reviewers did not believe that there was this mysterious thing that was so cool but that nobody had seen before,' Barrangou says happily. 'Or they thought it must be a peculiarity specific to this organism that nobody should care about. These unknown authors in this food company, working on this mysterious system, it just doesn't pass the smell test. It's not believable, it's like a fairytale.'

Then in 2012, almost overnight, everything changed. That was the year that biochemists Jennifer Doudna and Emmanuelle Charpentier revealed in the journal *Nature* how the amazing genetic sequences Mojica had found, and that Barrangou had proved was an immune system, could be repurposed as an incredibly powerful genetic engineering tool. So powerful, in fact, that Doudna has since revealed she has nightmares about giving the technology to Hitler.[80]

The key to the development of the CRISPR system into a programmable gene-editing technology is a protein called Cas-9 – 'Cas' being short for 'CRISPR-associated-protein'. Researchers had found that there were several Cas proteins in bacteria and archaea, all performing different roles in the coordination of this bacterial memory and defence system. Cas-9 was particularly interesting – its role was to take a sequence of phage DNA that the bacteria has 'memorised' into its own genome, and then roam around the cell scanning for DNA sequences that match. When it encounters the exact sequence, it chops it in half – and the invading phage DNA is effectively destroyed.

Doudna, a biochemist rather than a microbiologist, wondered if this might be something far bigger than a way to stop phages from spoiling yoghurt. In the great tradition of molecular biology, she was immediately interested in how she could use these biological scissors as a tool in her lab. What if, she wondered, instead of scanning cells for phage DNA, she could get Cas-9 to scan cells for other types of DNA? It was one of Doudna's researchers, Martin Jinek, who made the incredible discovery that Cas-9 could be 'programmed' to target any sequence of DNA you like.* You simply need to feed it a strand of 'guide' RNA (the chemical cousin of DNA) that matches the DNA sequence of interest, and off it goes, accosting every molecule it comes across until it finds a matching sequence. Then it very precisely chops it in half.

The programmer can pinpoint the exact letter in a DNA sequence billions of letters long where Cas-9 will make its cut. If a piece of DNA is added with ends that match the two ends that have been cut apart, there is a good chance the DNA strand will stitch that into the sequence as it repairs itself. Voilà,

---

* While Doudna and Jinek's paper was the first published work on how to use CRISPR as a gene-editing tool, a rival research group at Harvard's Broad Institute, led by Feng Zhang, fast-tracked a patent for CRISPR gene-editing methods soon after. Theirs was focused on using CRISPR gene editing to modify the cells of *higher organisms* like mammals and was processed and granted before Doudna's patent. An extremely complex and long-running dispute over who owns the rights and royalties of CRISPR technology has ensued. Although Doudna and Charpentier were recognised for their pioneering work with a Nobel Prize in 2020, as of 2022 the legal battle is still ongoing.

the Cas-9 has precisely found, chopped out and replaced a single gene, all at once. What's more, this whole incredible genetic system can be added into other cells – plant, animal, mammalian – and human, too – and used to target any DNA sequence in them that scientists could ever want to change.

CRISPR technology has made genetic engineering better – more precise, far simpler, far cheaper, but also more versatile. Just while writing this chapter, I've read about CRISPR being used to help reverse a genetic trait that causes high cholesterol; to breed faster race horses in Argentina; to make pig organs more compatible for donation into humans; to make new treatments for cancer and genetic diseases; and even to try and make the woolly mammoth 'de-extinct' by patching mammoth DNA into an elephant embryo. Scientists have made tiny versions of Neanderthal brains using CRISPR; they have 'corrected' the gene that causes sickle cell disease in patients with the disease and are hard at work testing many other types of CRISPR gene-editing medicines for other types of genetic disease.

It is generally seen as the great scientific development of our century so far, already finding uses in all sorts of areas across the life sciences and opening up the world of genetic engineering to labs and people who wouldn't normally be able to afford such technology. With individual genetic changes now made easy, scientists can now stack up lots of genetic changes together to reprogramme cells in amazingly complex ways. The myriad functions of genes found in nature can be combined, like components on a circuit board, to form new and exciting biological systems, a field known as synthetic biology.

Most biologists believe that it is inevitable that CRISPR will eventually lead to people altering their own genes, then those of their children, first in ways that help to eradicate genetic diseases and then in ways that will change humanity more fundamentally.

Scientists have already begun to test the safety and efficiency of editing human embryos to remove disease-causing genes, both with and without the approval of the scientific community,[81] and amateur scientists have begun experimenting with 'CRISPRing' themselves in unregulated and sometimes dangerous ways. One, Josiah Zayner, live-streamed himself injecting CRISPR compounds that he had modified to supposedly turn off the genes that limit muscle growth,[82] a stunt straight out of the pages of a Marvel comic (it didn't seem to work, and Zayner has now said he regrets trying to make himself into the Incredible Hulk). Russian President Vladimir Putin has spoken of the possibility of making gene-edited soldiers who cannot feel pain.[83]

The development of so-called 'gene-drives' – special tweaks that help genetic modifications from the lab spread quickly throughout wild populations – increases the risk that genetic engineering could have unintended, uncontrolled and potentially calamitous consequences. This technology has already been used to make large wild populations of disease-carrying mosquitos infertile, causing their numbers to crash and helping reduce the number of infections in that location. Of course, more nefarious uses of this self-propagating technology are not hard to imagine, from self-spreading genes that help sabotage a nation's crops and food sources to potentially genocidal human gene drives.

The possibilities are so endless as to be dizzying. CRISPR has become a sort of shorthand for the power of modern biotechnology, so much so that it's easy to forget where it came from. It was not *invented* by scientists per se – but has been slowly evolving inside microorganisms for billions of years as a particularly sophisticated way to protect themselves from phages.

The discovery of CRISPR also has big implications for the use of phage therapy in the fight against bacterial infections. For many decades before defence systems like CRISPR were discovered, scientists and doctors were using phage therapy without the knowledge that many bacteria have this highly sophisticated anti-phage weaponry to hunt down and shred phages' DNA. No wonder phage therapy has been inconsistent over the years. But what if we could turn the bacteria's CRISPR systems off?

Professor Sylvain Moineau is another phage scientist who has been working on CRISPR since its early, unglamorous days, before it exploded into a revolutionary technology and became shorthand for cutting-edge genetic engineering. The strawberry-blond microbiologist is the current curator of the Felix d'Herelle Reference Center for Bacterial Viruses at the University of Laval in Quebec City, which maintains a frozen catalogue of the world's most unique and historic phages.

Moineau also first encountered CRISPR while trying to control phages in the food industry. He's used to people not knowing what a phage is, but he wants people to know this amazing technology that is all over the news came from a natural process happening all around us every day. Recalling his career from an armchair in rather lovely-sounding Franco-Canadian English, he says, 'I tell people, when they are eating yoghurt on their breakfast in the morning: "you are eating CRISPR Cas9."'

As well as helping manage 'master copies' of interesting or unique phages from scientists all around the world at the Felix d'Herelle collection, Moineau is now studying the clever ways phages can evade CRISPR when infecting bacteria. The goal is an almost complete reversal of early CRISPR work, where researchers wanted to help bacteria fend off phages: interest has now turned to how phages can beat the bacteria. 'If you want a phage that is effective at killing bacteria in a therapy, using one that has anti-CRISPR systems would be wise,' says Moineau.

For example, phages can deploy special proteins to inhibit key elements of the bacteria's CRISPR machinery, or to immediately repair bits of their DNA that have been detected and chopped up by CRISPR;[84] some studies have even found that different types of phages can co-operate to overcome bacteria with CRISPR immunity, with one phage blocking the CRISPR system while another one replicates.[85] In 2021, a group of researchers based at institutes in Paris, Berlin, Leuven and Pittsburgh discovered that some phages use an entirely different genetic code to the rest of life, swapping out one of the biochemical building blocks of their DNA with a new chemical subunit so that it cannot be recognised by its host's defences. Rather than their genetic code reading as a giant sequence of As, Gs, Cs and Ts, these phages' genomes are made up of the letters G, T, C and Z.[86]

Systems to outwit or suppress CRISPR's DNA-chopping abilities could be very useful in other contexts, too – for example, if scientists want to edit genes causing disease in certain cells in the body but not others, or if the effects of a gene edit need to be suddenly suppressed or turned on and off. Scientists have identified at least ninety different anti-CRISPR compounds, many of which could be of use as natural 'brakes' for CRISPR technologies, or in the fight against infections that are both drug-resistant and phage-resistant.

The unbelievable success of CRISPR as a gene-editing technology is changing the way people study phages, says Moineau. Only around 40% of bacteria use CRISPR to defend themselves against phage DNA, he says – and so logically 'there are likely to be other exciting defence systems bacteria use against phage attacks'. Phage biology – once a small and collaborative field – is now awash with dozens of commercial companies looking for interesting molecules and 'the next CRISPR'.

The genes involved in anti-phage defence systems are often clustered together in what are called 'defence islands' within the bacteria's genome. Given that the defence genes are all located together in the genome, scientists are able to assume that the many mysterious sequences found within these regions are new and unknown defence systems. As well as producing molecules that may one day power new biotechnology like CRISPR has, many of these defence systems are related to the ways in which multicellular organisms like plants, animals and even humans defend themselves from viruses. In fact, recent studies have shown that central components of our own cells' defence against viral attacks have their evolutionary roots in the ways bacteria protect themselves from phages.[87]

A better understanding of these systems and how they work could therefore be hugely important in the development of new antiviral drugs, urgently needed to treat diseases like COVID-19 and the next pandemic viruses of the future. For example, a group of diverse antiviral molecules known as 'viperins' are found across bacteria, archaea and humans, and studies have found human viperins can block phages from infecting bacteria.[88] If the opposite is true, and bacterial anti-virals can prevent viral infection in human cells, then we should perhaps start to think of bacterial phage defences as an enormous repository of potential antiviral drugs. Given the vast number of unstudied or unknown viruses out there, and the never-ending churn of new innovations and counter innovations, nature's 'dark matter' will almost certainly contain other molecular marvels that we can use to master our genome or the genomes of our microbial enemies.

In the intricate lives of bacteria and their viruses, there are layers of complexity and mystery being found that the early phage scientists could never have imagined. In 2012, just before Jennifer Doudna and Emmanuelle Charpentier first announced they had turned the CRISPR bacterial immune system into a world-changing gene-editing tool, another researcher at Berkeley named Kim Seed was exploring the bacteria that causes cholera, *Vibrio cholerae*. The interaction between *Vibrio cholerae* and phages is particularly interesting: many strains of *Vibrio* are common marine microbes and not dangerous to humans at all; it is only when they are infected with certain phages that they become nasty pathogens.

Seed was interrogating the genomes of these phages in the hope of better understanding the dynamics between phage and host in the guts of cholera patients. After sticking some DNA sequences into the same online database as Mojica had done all those years ago, she found something odd: these phages appeared to have CRISPR-like DNA sequences too, just like their bacterial hosts.

'I was like "OK, that doesn't belong in a phage genome, that makes no sense",' admits Seed, a young associate professor, with big rectangular glasses and a business-like smartness. The phage's CRISPR DNA sequence was shorter and simpler than those in bacterial and archaeal cells, but was 'a bona-fide CRISPR system', says Seed, with bits of phage DNA that had been chopped up and added into the phage's genome as a kind of memory of past battles. Seed had no idea why a phage would have its own 'anti-phage' immune system.

Research has since revealed that some phages use CRISPR to chop up the DNA of rival phages that have infected the same cell. They have repurposed an anti-phage system to help them eliminate their viral competition. But what Seed discovered her phages did with their CRISPR system is even more mind-boggling. The phage was protecting itself from an even simpler parasite, also living in the same host cell. Known as a 'chromosomal island', these tiny bunches of genes have evolved to selfishly self-propagate themselves by jumping from cell to cell aboard phages when they replicate inside a host. In other words, they behave like viruses of viruses.

The parasites somehow insert themselves into a newly created phage's head in place of the phage's DNA, so that when the new phage particles go off to infect other cells, they simply inject the chromosomal island into a new host instead. This of course kills off the phages' chances of replicating – they are now instead just floating around infecting bacteria with more chromosomal islands. (If it turns out there is a virus of the virus's virus too, I give up.)

This fascinating and complex battle for replication rights, raging within a single bacterial cell, has surprisingly big implications for us up here at the human scale. Without its CRISPR system, the phage would not be able to replicate within *Vibrio cholerae* because of its parasite, the chromosomal island. Without the phage replicating, there would be fewer *Vibrio cholerae* turned into toxic pathogens. That would mean fewer pathogenic bacteria out there waiting to make humans very sick with cholera. The dynamic battle between the virus, its host and the virus of the virus drive the evolution of new strains of toxic cholera each year around the world.

What's even more exciting is to imagine what such CRISPR-armed phages can do for us. Firstly, having phages that can overcome viral parasites is useful when developing phage-based therapies. 'These parasites can block phages, and so if we know the bacterium that we're trying to target has one of these, we can arm the phage to be able to overcome them,' says Seed.

Phages armed with CRISPR systems could even be used to chop out the genes within bacteria that give them resistance to antibacterial drugs. For example, a patient with a drug-resistant infection is treated with a phage that not only infects it and kills it, but also chops up the bacteria's drug-resistance genes. Antibiotic drugs, now effective once again, are delivered to the patient as a final howitzer on whatever bacteria are left. It's insanely smart twenty-first-century medical technology.

'Of course, nature already came up with the idea,' Seed reminds me with a smile.

With more amazing news about the uses of CRISPR appearing in the science press almost daily, Mojica is still in Alicante, still looking at weird and obscure microorganisms scooped out of the hot and beautiful environments around him. Except nowadays, he is looking for viruses, too. His team casts the net wide, taking large samples of water, or soil, or whatever they can get their hands on, and running an analysis on all the morass of different types of DNA that can be found in there all together. When they find a sequence that seems like a CRISPR structure but is a little different, they zoom in and investigate. 'We are looking for peculiarities, or CRISPR systems that are different from what has been described already,' he explains.

CRISPR has helped spark a resurgence of interest in phages in molecular biology, and has inspired a new generation of scientists wanting to study how microbes defend themselves against phage attacks, says Mojica. He believes that phages have been overlooked in the history of microbiology so often because they are seen as mere 'genetic elements' rather than sophisticated living things. He hopes that this view is starting to change, and that people begin to understand that phages and viruses generally are a fundamental element of the puzzle of life, including us.

'Phages were underappreciated, for sure. But the situation has changed. People in my lab, in my department, understand microbial ecology. We are moving towards the field of viruses; people are interested in defence systems against viruses. Up until 2005 there were just a handful of antiviral defence systems known, and now we think there could be fifty, at least. People are getting interested in that, and also getting word of the relevance of viruses in general for the evolution of the host. They are the main driving force of evolution.'

Despite their reputation as harbingers of death and disease, all types of living cells really need viruses, Mojica tells me. 'Yes, sometimes the viruses kill the cell. But also cells get a lot of information, very crucial information from viruses. I think people are now becoming aware of that.'

The decades-long story of the discovery and development of CRISPR gene-editing is just another example of the importance of studying and understanding phages, and how work that can initially appear obscure and arcane can lead to the most dramatic advances. We have already gained so much understanding and technology from studying just a tiny fraction of the phages that exist in nature. It now raises the question – what else is out there and how do we find it?

# 15

## Find your own phage

In the spring of 2020, ten-year-old Isaac Temperton set out with his dad from their home in rural Devon, southern England, for their daily walk. That spring, the SARS-CoV-2 virus had spread so voraciously among the UK population that the government had been forced to impose a once unthinkable national lockdown. The hour of outdoor recreation allowed under the government's emergency coronavirus regulations had made a walk in the local woods or park an important, if oddly repetitive, feature of lockdown life. But today, Isaac was tasked with a special mission: to get a sample of water from the Lemon, a tiny river that ran through the woodlands behind their village. Amid that worrying global news about a new viral pathogen, Isaac's father, Dr Ben Temperton, a microbiologist at the nearby Exeter University, was hoping his son could help him find some good viruses.

Under the dappled shade of summer foliage, Isaac chose his spot. He carefully submerged a few small jars under the clear and crisply cold water, screwed on the lids, and headed back up the muddy bank with his dad to preserve anything they'd found from the river in the family fridge. The following Monday, his father took the water samples to his microbiology lab, which sits on the top floor of a large concrete tower rising from the steep hills of the university campus. He spun the samples in a centrifuge to remove any lumps of soil or vegetation, then passed the water through progressively smaller filters to remove anything larger than a virus – pretty much all other microbes. He poured the filtered water over various square Petri dishes covered in different types of bacteria, leaving them to incubate overnight. If, in the morning, any of the uniform lawns of bacteria on the dishes had spots or holes in them, Temperton would know he'd found a virus that could infect and kill that particular strain of bacteria.

Originally a bioinformatician, then a marine microbiologist, Temperton's attention had slowly turned to phages and their potential role in countering antibiotic resistance. Keen to find viruses that could kill the most common drug-resistant bacteria causing havoc in hospitals around the world, he was exploring whether ordinary citizens – or even ten-year-old kids – could go around helping him find such phages in their local environments.

The next day, examining a batch of plates, he found one of the samples taken by his son did indeed have plaques. Unlike the red spiky blob splashed across newspapers and emergency news broadcasts, he and his son had found a good virus, bobbing in the water of a local stream. And the type of bacteria Isaac's virus has evolved to seek out and destroy in particular happened to be one called carbapenem-resistant *Acinetobacter baumannii*, or CRAB. It's listed by the World Health Organization as the number one most concerning drug-resistant bacteria on the planet.

A 2017 study by the UK's Office for Health Economics estimated the cost of developing a new antibiotic to be around $1.5 billion.[89] Just the exploratory phases of research, before the efficacy and safety of a compound is tested in clinical trials, can take an average of four to five years, with another five years on top before the drug is tested and brought to market. In Devon, a ten-year-old boy found a voracious killer of the world's most dangerous bacteria using little more than a glass jar. What's more, as we ingest billions of them in our food, drink and other activities every day, these organisms are generally considered non-toxic by regulators like the US FDA. But getting such potentially powerful phages into clinical use remains complex.

Temperton admits that the idea of using phages in medicine is still so unusual in the UK that he and his colleagues weren't really sure where to begin, other than to keep finding and collecting as many of these viruses as possible. His phage-hunting forays in the Devon countryside with his son have grown into a project called the Citizen Phage Library, which asks students and members of the public to go out and find their own phages in their local environments. After collecting water samples with a simple kit, the citizen phage hunters simply log the date and geolocation of each one on the project's website[90] and post them to Temperton's lab in Exeter for processing. The samples are screened against a panel of drug-resistant bacteria currently posing the greatest risks to human health. Once Temperton's lab confirms a phage has been found in a water sample, the person who sent it in gets to name their phage (or phages) and receives an electron microscope image of it. They've found hundreds of potentially very important phages already in this way.

Critically, the cost of isolating and processing each phage is low – a few hundred pounds – and it takes just a few weeks. As well as amassing a bank of phages that may come in useful locally or nationally, Temperton's broader aim is to create a template for open-source and low-cost phage libraries that can be replicated in countries where resources are limited. It's a win–win: local people get to have fun while learning about the importance of phages and making their own small contribution to science, and local scientists get a steady supply of potentially useful viruses to add to their growing banks of phages. Or that's the theory, at least. When I eagerly volunteer to take part and find a virus for the Citizen Phage Library, I discover the most abundant biological entity on Earth can be surprisingly elusive.

I arrive at Ben Temperton's lab with nine pots of pretty icky water. In preparation for my visit, I've been sent a pack of small glass jars to collect samples, and I've spent the week scooping water from a range of weird and dirty locations. I'm confident the dirty places I've sampled will be brimming with viral life – especially the murky canal behind London's King's Cross Station, and the green stagnant water from a cat bowl that has lain in my garden for months.

As I'm shown round Temperton's lab and introduced, he and his colleagues are still buzzing from a bumper crop of over fifty new phages found recently by a group of local students. One of his PhD students is processing particularly dirty-looking samples from a local community poultry farm known as 'the Chicken Collective'. When the steady West Country rain suddenly becomes a thundering downpour, one of the lab technicians runs out to sample the muddy rivulets streaming across the pavements outside. There is a feeling that phage discoveries are everywhere for the taking.

The process of 'isolating' a phage – that is, revealing and capturing a virus by seeing it destroy and replicate on a plate of host bacteria – is effectively unchanged from the method d'Herelle developed more than a hundred years ago in the 1910s, albeit with a slightly horrifying amount of single-use plastic rather than glass.* The simplicity of the method is exciting for the prospects of phage science in developing nations – you really can do this with the most basic microbiological supplies.

Today, for simplicity, I am just testing my samples to see if any contain phages that grow on *E.coli*. The pervasive sweet cheesy smell of the bacterial broth takes me back to my own undergraduate degree, now so long ago I might as well not have done it, and I am soon reminded that microbiology is a skilful vocation. As well as keeping track of the various containers at hand – my water samples, buffer solutions, bacterial broths and pure water all look virtually the same – one must constantly avoid contaminating the samples and equipment by contact with an unsterile surface. This challenging mix of physical dexterity and concentration is known as 'aseptic technique': lids of jars can't be put down on the bench for a second, so jars must be held and unscrewed with one hand while liquid

---

* And I mean lots – I had heard science has a single-use plastic problem, but it's depressing how many pipette tips, syringes, screw-cap tubes and other thick plastic items I bin immediately after using once – all of which come wrapped individually in plastic.

is pipetted in with the other. Fresh equipment must be used for every sample, used straight from the plastic wrapper without touching anything.

The thought of working like this without the convenience of single-use plastic boggles my mind, but d'Herelle was apparently a master, able even to handblow his own glass flasks for specific jobs. Just like d'Herelle and the microbiologists of old would have done, I work beside the flickering flame of a Bunsen burner to give a halo of semi-sterile air in which to work, an old-school touch in this hyper-modern lab.

My samples are centrifuged and filtered to remove any particles of dirt and bacteria. While we wait, I label up several tiny vials of amber glass, the deep orange colour to help protect the phages from UV light. Ironically, as the only writer in the lab, the challenge of scrawling key information in tiny writing on lots of identical vials is almost too much for me. I am soon hopelessly confused in a way that I remember well from university.

Each of my spun and filtered water samples are mixed with a solution of healthy, happy *E. coli* bacteria, and carefully poured onto a nutrient gel on square plastic plates. This should grow into a layer of hazy bacteria, and if any phages active against *E.coli* are present, within twenty-four hours clear holes or plaques will emerge from the haze, each one representing a tiny epidemic of phage infection and bacterial death.

My pile of plates is taped together and placed with several dozen others from the lab to incubate overnight at 37°C. I retire for the day to my seaside bed and breakfast, hoping that somewhere on my plates, at a microscopic level, all hell is breaking loose.

Generally, phages are pretty stable if the environment they are kept in is not hostile or markedly different to where they naturally reside, and researchers have been known to keep phages in their fridges for over forty years with no reduction in their concentration or activity. However, phages do tend to stick to plastic and can break down over time if not kept cool. They can also be broken down in UV light, or damaged by abrasion or exposure to chemicals. I begin to worry about my leisurely day spent sampling in London; my long drive down to Devon and the hours I spent with my water pots sloshing around in a bag on my back. Did I look after my little phages properly?

The next day, Temperton greets me in a low-key way, like a doctor with bad news. 'It's not looking like we've got any obvious plaques,' he says, the hint of hope preventing me from being too crushed. 'But we'll take a closer look under the lights.' We hold each one of the square plastic plates in front of a bright LED light that will accentuate even the faintest plaques. Each one is perfectly, uniformly milky – just *E.coli* bacteria growing happily. Something on the last one catches Temperton's eye – small, circular, see-through . . .

It's a bubble. We are both leaning over, squinting at a bubble.

Maybe I didn't get them in the fridge soon enough? Maybe the long drive down to Devon was too much for them? Or maybe I really messed up my lab work in some fundamental way? Who knows? Perhaps I managed to find the only six spots in England without any *E.coli* phages. Around me, though, there is excitement in the lab: a phage isolated from one of the Chicken Collective's coops has been found to kill a broad collection of *Pseudomonas aeruginosa* strains isolated from cystic fibrosis patients. What's more, the farmers have opted to

call this potentially life-saving phage *KylieMinegg*, which everyone agrees is one of the best names yet. But I'm gutted. My
search for a phage I can call my own continues.

Exeter's Citizen Phage Library is just one of several 'phage-
hunting' projects gaining momentum around the world to help
enlist students and members of the public to find new and
potentially useful phages. The largest is probably the
SEA-PHAGES programme,[91] devised by Professor Graham
Hatfull, a phage expert at the University of Pittsburgh. Through
the programme, over 5,500 students at over 170 universities
around the world have been shown how to find, isolate,
sequence and name their own phage. This has helped add
20,000 phages to the world's database of known viruses, over
half of which are relevant to Hatfull's particular interest,
*Mycobacterium*, the bacterial group behind the deadly disease
tuberculosis (TB) and other nasty infections, including leprosy.

Again, the programme is a win-win for everybody involved:
Hatfull gets to learn more about the diversity and abundance
of *Mycobacterium* phages in different corners of the world;
young students from a diverse range of backgrounds get to try
real lab work and experience the thrill of scientific discovery
long before most scientists do in their careers, and the entire
project fosters an excitement and interest in the amazing viruses
that can be found in the most mundane places.

'Soil and compost are places where you're likely to have the
most success in isolating *Mycobacterium* phages, but I think that
students have probably looked pretty much anywhere and
everywhere that you can imagine,' Hatfull tells me from a huge

book-lined office in Pittsburg. 'They can be pretty creative. Students have looked in snow, and lakes, and detritus from gutters, and all sorts of other places.'

It was the SEA-PHAGES programme that helped South African university student Lilli Holst find phage 'Muddy' on the underside of a rotting aubergine in her parents' compost heap. Muddy, if you'll recall, became one of three phages used in an experimental treatment that helped control a terrible drug-resistant *Mycobacterium* infection that had spread throughout the body of a seventeen-year-old British girl following a double-lung transplant. Hatfull was contacted by the patient's mother and set to work testing phages in his collection against the young girl's specific bacterial strain from his huge collection of thousands of phages. Just three phages infected the strain, including Muddy, the phage found in Durban almost a decade earlier. Another phage called ZoeJ had been found by a student in soil outside the science labs at Providence College in the East Coast state of Rhode Island, and a third, called BPs, had been found by an undergraduate student at the University of Pittsburgh. After ZoeJ was genetically modified to be more violent – it had tended to lurk inside the bacteria rather than bursting it open – the three phages were shipped across the Atlantic to London's Great Ormond Street Hospital.

'To hear it was being used ten years later for this really specific strain of bacteria, it was just bizarre,' says Holst, who now works for a communications company. 'Exciting and bizarre. My parents are very proud of what came out of their compost heap.'

Hatfull now describes Muddy as 'probably my favourite phage to use therapeutically'. The virus from the rotting aubergine in Durban has since been deployed in emergency phage therapy cases more than a dozen times, including in a severely immunocompromised patient whose arm had been partially decomposed by a horrendous drug-resistant strain of the bacteria *Mycobacterium chelonae*.[92]

Not all students are as lucky as Holst in finding a phage like Muddy. Most phages found by hunters will simply sit in a fridge and be added to the ever-growing open digital databases of viruses shared by researchers. That is still useful – these phages and their genomes can be used to analyse the relationships between different viruses and their hosts, helping us better understand how they interact and which genes do what. 'Students don't know what they're gonna get,' says Hatfull. 'They don't know if what they find is going to be useful or not. But the possibility is there. And just the mere possibility adds a degree of engagement and excitement for the students. Even if the particular phage that you discover does not get used therapeutically, which is going to be true for most of them, you've added a small but important brick in a very big wall.'

With new phages being found every day by professionals and amateurs alike, plus batches of thousands of new species being discovered all at once from metagenomic sampling, the important job of classifying and categorising all these viruses – known as viral taxonomy – has understandably become rather difficult. Given their propensity for swapping genes and evolving very rapidly, there has been some debate over whether classic concepts of biology, such as 'species', 'strain' and 'population', even apply to these life forms. They might perhaps be better thought of as a dynamic, ever-changing pool of parasitic genes that appear fleetingly in particular combinations. But even in such fast-evolving and hyper-diverse organisms, classification systems are important. Taxonomy helps ensure researchers have a shared language with which to describe, compare and study viruses.

Belgian virologist Dr Evelien Adriaenssens introduces herself to me by email as the person 'partially responsible for phage taxonomy not being easy to describe anymore', followed by a winking emoji that suggests she's just being glib. But she kind of is that person. Since 2020, she has been the chair of the Bacterial Viruses Subcommittee of the International Committee on Taxonomy of Viruses (ICTV), the global body that is in charge of naming viruses and classifying them into species, groups and families.

It was while working with viruses from the Namib Desert in Southern Africa that Adriaenssens realised that the system for classifying and naming phages was not fit for purpose. The desert stretches for over 2,000km and the rock and sandy soils here have been baked by the sun without regular rain for over 50 million years, making it the oldest desert in the world. The scraps of humidity that roll across this place from the Atlantic Ocean mean it is not quite as arid as South America's Atacama Desert, officially the driest place on Earth, but it's an inhospitable place: the desert's name translates, roughly, from the local Nama language, as 'the area where there is nothing'. However, the tiny amount of moisture here supports a surprising variety of life. Among the most intriguing are hypoliths – small communities of green life that form on the underside of tiny glassy stones and rocks. The minerals let through enough light for a tiny layer of photosynthetic microbes to eke out an existence beneath, protected from the harsh wind and UV light and exploiting the miniscule amounts of water that might become trapped here.

Although these hardy microbes have been studied extensively, nobody had ever thought to look for viruses among them, until Adriaenssens came to the Namib as a postdoctoral student on the hunt for scientific adventures. When she took some of the hot glassy rocks to analyse at the University of Pretoria, South Africa, she found dozens of different phages, and many more in the open, baked earth nearby. She believes the majority of phages here lie low within their host, integrated peacefully in the bacteria's genome, sometimes for years on end. When it finally rains, enlivening these tiny ecosystems with new life and growth, the phages switch to the more violent lytic strategy, suddenly replicating and bursting out of their hosts and into the quenched soil to find new hosts.

When Adriaenssens went to analyse what kind of phages were present in the desert, she found pretty much all the viruses in her samples were unclassified or unknown to science. 'That's when I really realised how much work there was to do. How can we understand what's going on in, say, the soil, or in your gut, if we don't have the words and terminology to actually describe what is there?'

Before the invention of DNA-based technology like metagenomics allowed scientists to 'find' tens of thousands of viruses at a time and compare their DNA sequences, phages were classified by looking at the similarities or differences in their visual appearance under a microscope. The godfather of phage taxonomy, Hans Ackermann, published over 230 articles and five books describing the shape, size and characteristics of phages and other viruses, assembling what may be the largest bacteriophage collection in existence* and estimated that he had personally investigated almost 2,000 different phages.

'I like to collect stamps, phages, and airport stops,' he wrote in a reflection on his scientific career in 2012. 'I have examined more than 1,700–1,800 bacteriophages and intend to be the individual who has seen the most bacteriophages in an electron microscope.' His other passion was Haitian zombies.

With echoes of phage heroes past, Ackermann escaped East Germany using a borrowed passport and went on to a career at the Pasteur Institute before settling in Quebec. He became known for his extraordinary skill with an electron microscope and, as one of his peers tells me, for getting 'very, unreasonably upset' about the quality of others' images. He would examine each phage, magnified by tens or hundreds of thousands, describing their minute structures in minute detail – the shape of its baseplate, the number of striations on its tail, the configuration of the tiny tail fibres – placing phages with similar characteristics into groups. Ackermann's classification of phages according to their physical structure and appearance was used by microbiologists for decades. He was even honoured with his own group of viruses, the family Ackermannviridae.

However, in the early 2000s, virologists like Forest Rohwer and Mya Breitbart began to take a new approach, gathering samples containing thousands of viruses at a time and exploring

---

* The collection is now available through The Félix d'Hérelle Reference Center for Bacterial Viruses, at Laval University.

how different or similar phages' genetic sequences were. As reading genetic sequences became easier, cheaper and the go-to technology for understanding life on Earth, virologists began to realise that the divisions devised by Ackermann did not accurately reflect the true evolutionary relationships of these phages. Some phages that looked superficially similar had radically different genomes, and some, which looked quite different, were found to be surprisingly closely related in terms of the genes they possessed.* The system Ackermann had developed, with all his thousands of hours of study with the electron microscope, would, therefore, have to be completely rethought. For years since, Adriaenssens and a group of collaborators from around the world have been working on the long and difficult process of disentangling the older groups and the appearance-based classification system in favour of a genetics-based one.

Adriaenssens only met the fiery Ackermann once, and he was becoming ill as plans to change his life's work began to take shape. She describes him as 'critical but supportive' of efforts to change the system for classifying his beloved phages. 'And he wasn't afraid to be critical,' says Adriaenssens. 'Sometimes I would send him an electron microscope image and he would say "that is garbage and whoever took it needs to be shot".'

By 2021, when Adriaenssens and the rest of the Bacterial Viruses Subcommittee announced their proposals for new groups based on the genetic relatedness of phages,[93] including the abolition of several groups, there was a happy ending, of sorts. One of the newly identified groups of genetically similar

---

* Reminiscent of how organising people into several broad 'races', based on just our appearance, is actually very unhelpful in terms of understanding the diversity and genetic relatedness of different people around the world.

groups of phages seemed to match up fairly well with one of the older, visually similar groups, and so could stay largely unchanged. It was *Ackermannviridae*. 'In this case, Hans had actually accurately defined the genetically related family just with electron microscope images,' says Adriaenssens.

Away from the difficulties of classifying thousands and potentially millions of phages into logical and meaningful groups that everyone in your field agrees on, ever more people are enjoying the thrill of finding and naming their very own phage. Phages that have DNA sequences over 95% similar to one another are classified as being of the same species, and naming a species is still a matter for the Bacterial Viruses Subcommittee. But naming *individual* phages – that is, a phage found at a particular location with a unique DNA sequence – is much more anarchic and fun. And it's not at all scientific or serious.

The first rule of naming phages at the virus database PhagesDB.org is, for example, *'do not name your phage after Nicholas Cage'* – a tongue in cheek reference to the fact that too many people were using the same rhyming pun based on the Hollywood legend's name. The rules go on to advise phage hunters to *'think of choosing a phage's name as similar to choosing a child's or pet's name'* and to *'stay away from political, controversial or violent names'*. And that's pretty much it.

And so there are now phages out there with all manner of mad amusing names that, in a rather beautiful kind of way, reflect not just the diversity of the phages on this planet but the diversity of the people finding them too, from the sensible and orderly to weird and wacky, all the way through to infantile and puerile.[94] As the bacteriophage postdoctoral student Frank Santoriello put it on Twitter: 'Bacteriophages are either named something like phiKF892-1, or something like KidneyBean. There is no middle ground.'[95]

As student and citizen-led phage hunting efforts grow, we can expect more wonderful stories of people finding life-saving viruses in the most unlikely or mundane places – at the bottom of their garden or in muddy streams. It's truly exciting to think that there are literally trillions of powerful antibacterial agents out there – many with as yet unknown and mysterious features – just waiting to be found by people armed with little else but jam jars and a smartphone. And how wonderful to think that future antibiotic treatment regimens and serious clinical trial reports might be focused around viruses with names such as 'Frankenweenie', 'OneDirection' or 'KimJongPhill'. Or my personal favourite, found by a teenager no doubt – 'Whatever'.

# PART 5

*Future phages*

# 16

## Phage therapy 2.0

After over a hundred years of highs and lows, there is excitement about the power and potential of phages once again. New journals and scientific conferences are appearing focused solely on phage research. Blockbuster funding grants are being awarded for fundamental studies. Groundbreaking clinical trials of phage therapy are finally underway, due to report their results in the coming years, and each week seems to bring new headlines about successful experimental treatments. The global phage therapy market is now estimated to be worth over a billion dollars, and phage-based medicine, once derided as an idea for cranks and commies, is suddenly looking like quite a contested space – with many different approaches being explored, dozens of different companies emerging and money pouring in from both public and private sources.

But on the big question – of whether such treatments can ever be made affordable and widely available – the same old difficulties remain.

Back in the 1930s, Felix d'Herelle saw how phage therapy – requiring complex administration and infrastructure and not necessarily generating profitable products – had not been a good fit with companies chasing profits and healthcare systems looking for quick fixes. The same problems still hold today.

Putting phages into patients' arms still involves too much lab work, too much improvisation and too much guesswork. Commercialising phage therapy remains mired in technical, safety, logistical and regulatory challenges and, under the current system of emergency exemptions and small experimental projects, just a tiny minority of drug-resistant infections will ever be treated with phages. With the horrors of a post-antibiotic future now bleeding into the present, radical new approaches are needed.

Thankfully, a new generation of phage innovators are now combining the lessons of the past with cutting-edge technology to drag this one-hundred-year-old form of medicine into the twenty-first century. They are questioning some of the most fundamental ideas about what a phage *is* and what it can do. This is phage therapy 2.0.

What if, for example, instead of embarking on a laborious, time-consuming and sometimes intercontinental search for suitable phages for every patient, doctors could use a machine to simply create the exact phage for the job, there and then at the hospital or clinic? The idea is not as far-fetched as it sounds. In fact, as I write, the race is on to become the first group to treat a patient with a fully 'synthetic' phage – that is, a virus that has been made from scratch from chemicals, machines and pipettes, rather than from viral material replicating inside a bacterial cell.

One of those groups is led by Jean-Paul Pirnay, a researcher at the Queen Astrid Military Hospital in Brussels. Back in 2020, Pirnay took the unusual step of submitting what is essentially a piece of creative writing to the scientific journal *Frontiers in Microbiology*, entitled 'Phage Therapy in the Year 2035', which presented his high-tech vision for how twenty-first-century technology could help phage therapy shed its historical baggage.[1]

The article, now viewed tens of thousands of times, follows the fictional story of a retired microbiologist, John Iverian, who gets bitten by an insect while soaking in the bath in his Antwerp apartment. After realising the bite is infected with a dangerous flesh-eating bacteria, Iverian pulls out a futuristic black box known as a 'Phage-BEAM device'. A hologram called Marcia pops out to help guide Iverian through the process of taking a sample from his infected wound, before the device sequences the DNA of the bacteria, determines the best phage to kill that bacteria and then synthesises that phage there and then in his bathroom.

Despite sounding wildly futuristic, all the scientific technology in the Phage-BEAM device exists already – albeit not quite in a form that can be shrunk into a convenient black box that sits in people's bathroom cupboards. Extracting and sequencing DNA from samples is now so commonplace that it can be done in the middle of the jungle with a portable kit that fits in a rucksack. Several research groups have developed AI and deep-learning platforms that can predict what host a given phage will infect, based just on its DNA, finding potential matches in minutes rather than days or weeks.[2] And the first phage made with DNA synthesised chemically in a lab was made over twenty years ago.[3] The least plausible part in Pirnay's vision of the future is probably the hologram called Marcia.

Pirnay began working on phage therapy in the early 2000s, when the funders who rejected his first proposal ironically described it as 'science fiction'. He looks how you might imagine someone who writes bacteriophage fan-fiction might:

purple round-framed spectacles, an almost stripy black and grey beard and loud psychedelic shirts decorated with video-game characters, mushrooms and sometimes viruses, of course. In his home hangs a large and brightly coloured retablo – a traditional Catholic devotional painting – commissioned from an artist in Mexico for good luck before he first used phages in surgery. It shows surgeons operating over a prone patient, with a big orange thought bubble full of phages emerging from one of the surgeons' minds, and the Virgin Mary floating in the surgical lights.

Since he commissioned the work, he has helped treat over one hundred patients with phage therapy, and the phages he and his colleagues have helped match to patients have been used in thirty-five different hospitals in twelve different countries, from the US to China, Scotland to Tunisia. He has coordinated phage therapy for soldiers and civilians with war-like wounds in Belgium for around a decade, including a casualty from the Brussels Airport suicide bombing of 2016, whose numerous skin grafts and shrapnel wounds had become hopelessly infected, and an African politician who arrived with infected gunshot wounds. Pirnay has also been instrumental in persuading the Belgian health authorities to regulate phages separately from the way other drugs are regulated across the European Union, making Belgium a sudden hotbed for interesting phage therapy cases, trials and research, and a model for how phages might be regulated in other countries.

Despite his successes, he is realistic about what can and can't be done with phage therapy in its current form. The current process of sending patient samples off to labs around the world for matches – which must then be detoxified and purified into a pharmaceutical-grade preparation while the patient is in dire straits – is simply too time-consuming and labour-intensive ever to work as a mainstream form of medicine.

'A lot of people are talking about making great biobanks full of phages, and bringing in industrial partners to produce phages,' he comments. 'But you will still get bottlenecks. Not everyone will want to give their phages to public biobanks, and it is difficult to get industrial partners to invest in cases where a personalised approach is needed. I want to invest in the approach where any phage you want or need is just on a USB stick.'

The promise of synthetic phages is clear. As well as being manufactured when and where they are needed from just a digital DNA sequence, they could be designed to suit the exact requirements at hand. They could be made to be identical to ones that already exist on the other side of the world or designed from scratch, with a mix of useful characteristics. In contrast to naturally occurring phages, which normally contain dozens of genes which we know little about, synthetic viruses could be extremely minimalist, with only the genes absolutely necessary to build a virus that can infect and kill the bacteria at hand. Knowing exactly what genes are in each phage, and exactly what they do, could in theory make them more predictable and safer to use – and crucially, easier to license as therapeutic agents.

They could be loaded with genes to provide a range of add-on features, such as genes that help counteract the bacteria's defences or that counter the emergence of resistance; that help break down the tough 'biofilms' that grow around and protect bacterial cells or that resensitise bacteria to the antibiotics that they have become resistant to. Phages could also be designed to survive longer in certain parts of the body, be cleared more quickly, or replicate in a way that is most useful to a particular form of treatment or infection. Of course, building a viable living organism is more complex than just compiling a list of the genes you want, and it's unclear how the combinations of genes in a 'designed' phage would affect its ability to infect and replicate in a host.

For now, Pirnay is keeping things simple, and creating synthetic phages that are identical to phage strains found in nature. While his vision for a hologram-emitting, phage-synthesising black box might be some years away, he says he is close to treating his first patient with a synthetic version of a T7 *E.coli* phage, an audacious leap forward for phage therapy that will surely catapult it into the headlines once again.

Pirnay is clear that for a phage to be classed as truly 'synthetic', there must be no bacteria involved in any stage of the production process. That way, concerns about contamination with bacterial toxins or dangerous gene swapping during the production process can be put aside. Previously, some viruses have been called 'synthetic' because their DNA sequence was synthesised in laboratory, but the viruses were still produced by a bacteria 'infected' with the synthetic DNA. Instead, Pirnay and colleagues are able to produce synthetic phages using a completely cell-free system known as Phactory.*

This stunning process, developed by researchers at the Technical University of Munich, involves chemically synthesising the phage

---

* Phactory began life as an entry to the International Genetically Engineered Machine competition, or iGEM. This remarkable biology conference involves hundreds of groups of students competing to create the most innovative biological system from a standard inventory of cells, genes and proteins, known as 'bio-bricks'. It's like a giant Lego competition but for building new life forms.

DNA sequence, and then in a series of complex chemical steps, using that sequence to produce and assemble the protein parts of the viruses, replicating what would happen inside an infected bacterial cell. Pirnay describes it as being like a 'printer, or espresso maker' for viruses – but like so much of molecular biology, the reality is that to the casual observer it looks like a series of clear liquids being transferred into different vials.

It's no mean feat, conjuring up a new living thing from just chemicals. Some highly repetitive sections of the T7 genome have proven trickier to synthesise than others, delaying the project. But Pirnay is so close to creating a batch of synthetic phages he can use on a patient that he is already starting to discuss them with the relevant health authorities and regulators in Belgium. He's confident they will not be fazed by the fact these are effectively lab-made viruses – a scary term for a world awash with rumours about where COVID-19 really came from. 'I think the regulators will not care if the phages we use are natural or synthetic if they cannot tell the difference,' he says casually. 'To them, they are identical products.'

Painstakingly recreating something that grows so abundantly in the natural world may be just another example of human-kind's folly – or maybe it's the latest example of our increasingly impressive mastery of the inner workings of Earthly life. Either way, when that first patient is injected with a synthetic virus it will be a thrilling world-first. Could phages 'printed' on-site there and then finally transform phage therapy into a convenient and cost-effective form of medicine? Pirnay believes that his group's phage-making technology will soon interest the world's giant medical device manufacturers. A subscription-style product, providing hospitals with access to the software that determines which phage to 'print' from a database of phage DNA sequences, could be an even more attractive product for potential investors, he says. For the moment, it remains just another one of Pirnay's visions of the future. But soon, it seems, his science fiction will start to become science fact.

Pirnay is not the only one synthesising phages. Felix Biotech operates from a suite of offices and boxy labs in San Francisco's Bay Area, a sprawling stretch of the West Coast famous for its hundreds of tech start-ups. The labs may be small, but here they are working on a big idea: to design and build phages from scratch that can infect many different strains of bacteria at once.

The company's founder, Rob McBride, greets me on a call as you might expect the founder of a Bay Area biotech start-up to – casually dressed, with white AirPods in his ears, he crams in our video call between other virtual meetings as he strides between offices, labs and plush communal areas. The South African-born biotech entrepreneur founded Felix – yes, named after d'Herelle – on the principle that the company must overcome the main challenge that has prevented phage therapy from going mainstream in the past: mainly, that most phages only infect very specific strains of bacteria.

To counter these problems, Felix is developing a 'digital phage platform' – a kind of AI and robot-aided kit for designing and manufacturing phages, again without the need for bacterial cells. The company uses machine learning to take masses of information about existing phages and their characteristics and generate hypothetical new ones with the features they want. As well as broadening their phages' host range, for example, they may also include the genes that give some phages a talent for breaking down bacterial biofilms.

The team are also identifying combinations of genes that could be added to phages to try and delay or counteract the ways bacteria fight back against phages, preventing the development of resistance. Clinical trials of their most promising 'asset' so far are underway, and while they inch through these small, expensive studies, they are deploying their phages in various animal and non-medical scenarios – including pet care and personal hygiene products – to help generate more data to feed into the platform's algorithm.

Creating phages with a very broad range of hosts – ending the need to match phages to the patient's exact strain – is a more scalable and, crucially, *profitable* model than trying to manufacture phages that might work on just a few patients each, says McBride. If Felix can, for example, develop a phage that can treat virtually all strains of *Pseudomonas aeruginosa* – the cause of a vast range of infections, like those associated with cystic fibrosis, chronic obstructive pulmonary disease (COPD) and surgical wounds – it could be worth billions. Several new groups are taking similar high-tech approaches. Locus Biosciences, for example, backed by £80million of funding from Johnson & Johnson, has combined AI, automation and genetic engineering to create a cocktail of phages which they say can kill 95% of known strains of *E.coli* in initial tests.

Not everyone agrees that the world needs synthetic or genetically engineered viruses to develop phage therapy into a twenty-first-century medicine. In Gaithersburg, Maryland, a flat and spacious suburb on the far outskirts of Washington DC, long-time phage therapy advocate Carl Merril and his son, Greg, are building a different vision for how phage therapy might one day operate. On a wide, tree-lined avenue in Gaithersburg's business district is the headquarters of Adaptive Phage Therapeutics (APT), a long three-storey brick building adorned with the company's name and phage-like logo. Almost one hundred people work here, although the elder Merril, now in his eighties, says there seem to be more robots in the labs than people. The unit is so large he needs a Segway when he visits, he jokes.

Rather than tinker with phage genomes or produce them artificially, APT plans to apply cutting-edge automation to a more traditional approach that uses the virtually infinite supply of phages found in nature. They are building a giant bank of phages that infect common bacterial strains, sourced from dirty places and hospitals around the world, screening them for any dangerous genes and then purifying them ready for use. But rather than asking the FDA to approve each individual phage product when it is needed, the company is working to get FDA approval for their whole operation, from the way phages are isolated, screened and matched to the manufacturing process and treatment. If successful, it would essentially mean all the tens of thousands of phages in their biobank would be pre-approved for therapeutic use at once and ready to use.

Merril is disdainful of those attempting to genetically engineer or synthesise the perfect phage for their patients. 'Why would you want to do that?' he grumbles. 'You think that you're going to come up with something that nature hasn't already thought to try in four billion years of evolution? I don't think so.'

APT's headquarters is strategically placed to be within 50

miles of several major hospitals in Washington DC, and Maryland, and the company sends phages farther afield across the US. But the company wants to move on from shipping their products out on ice via couriers and helicopters. Merril Junior, a shaven-headed former race-car driver turned CEO, is helping develop what are essentially giant vending machines that they hope will be used to stock thousands of APT's pre-approved phages in hospitals around the country. When the labs in Gaithersburg find a good match for a patient in their phage library, they can simply send a code to the hospital treating the patient and the appropriate FDA-approved phages would pop out of the vending machine. If or when resistance to that phage emerges, ATP says they can just find a new one.

'With antibiotics, when you have resistance you have maybe a dozen other drugs you can try, at most,' says Merril. 'With phages you have 10 to the power of 31' (the preposterously massive total number of phages out in the world). 'There's practically an endless supply of them out there.'

To streamline the process, anything that can be automated has been automated – from the initial tests to see which of their phages work on patients' samples, to the purification and packaging of the final medicine. A state-of-the-art robot can fill thousands of completely sterile vials with a ready-to-administer dose of phage to pharmaceutical-grade levels of purity. It then sterilises itself before doing it again with a different phage, replacing a laborious clean-room process that could take days by hand. APT say they can provide a precisely matched pharmaceutical-grade product within twenty-four hours of receiving a sample of a patient's bacterial strain.

With this industrialised approach, APT is increasingly confident that they have the phages needed to tackle whatever dangerous bugs are circulating in the US and beyond. So confident, in fact, that in late 2022 the company launched a challenge to the infectious disease community: they will pay $1,000 to any researcher who has a strain of bacteria (from the CDC's six most dangerous bacterial species, the so-called ESKAPE pathogens) that cannot be killed by phages in the APT library. CEO Greg released an eye-catching statement that he is 'confident that APT has solved the problem caused by bacteria evolving resistance to antibiotics'. It's quite a claim.

But all this innovation and automation comes at a cost, and APT's bespoke pharmaceuticals are unlikely to be cheap. Merril Senior argues that in many cases the costs will be small relative to the huge ongoing cost of treating a drug-resistant infection, especially in problems like bone and joint infections following surgery. 'Take a hip replacement,' he tells me. 'I mean, they charge patients a lot of money for that. And if they have to take it out, it costs even more. Then if they have to amputate the arm or leg, you're talking a tremendous amount. So even though we may charge, you know, $20,000 or something for a treatment with phages, it's a bargain versus what they're going to face the other way.'

Still – it doesn't strike me as anything like a replacement for the convenience and cost of a packet of antibiotics.

The different approaches to phage therapy emerging across the world right now can be thought of as sitting on different positions on several axes – from those using naturally occurring viruses to those making new ones; from treatments that might work on many people with similar infections to highly personalised patient-specific treatments; from publicly funded and collaborative projects to private, patented and presumably expensive commercial ventures.

An oft-cited statistic is that by 2050, at least ten million people will die each year from antibiotic-resistant infections. But what is less well publicised is that up to 90% of those deaths are predicted to occur in Africa and Asia, where antibiotic use is soaring and access to alternatives is poor. So, while investment is now pouring into 'phage therapy 2.0', via glamorous West Coast start-ups and gleaming European teaching hospitals, the focus of our efforts in the war against deadly bacteria should arguably be focused on approaches that can deliver for the resource-poor countries in the global south.

With much of the world's phage expertise residing in Europe, the former Soviet Union or North America, a project known as Phages for Global Health has been set up to send some of the world's leading phage scientists to African and Asian countries, with the aim of inspiring a generation of new phage researchers across these vast continents. In intense practical courses, sometimes as short as just two weeks, local students and scientists learn the basics of phage discovery and purification. Since 2017 the programme has created groups of phage hunters in Uganda, Kenya, Rwanda, Nigeria, Tanzania, Ghana, the Gambia, Malaysia and Indonesia. Alumni from the courses are now holding annual training sessions to pass on their passion and skill for phage-finding to others.

In total, over 1,200 scientists have been trained in basic phage biology, hundreds of new phages have been discovered and 50 phage projects are underway. In Kenya, phage researchers have been providing phages to the hectic chicken farms and streetside butchers of Nairobi, helping to reduce food poisoning by *Campylobacter* bacteria; in the Democratic Republic of the Congo (DRC), a collection of phages that can act on a broad range of cholera bacteria is being used to decontaminate latrines and other water sources, as well as being given out as a preventative measure to those who have had close contact with people with the disease.

Phage therapies matched to individual patients and requiring high-tech equipment may never be economically or logistically viable in low-income countries. But more democratic, public health-focused approaches like those being explored in Kenya and the DRC could be. As Ugandan phage researcher Deus Kamya told me: 'We have always been just getting formulas for vaccines and antibiotics that the Western world has developed. But here comes phage technology – now we have a chance to really organise ourselves and set up a structure where we could harness the power of phages. Our sewage, our waters, the oceans – they are very rich sources of these phages, and isolating them, as we know, is not that difficult.'

Phage treatments against diseases like TB and cholera are especially promising for the global south: TB phages tend to have broad host ranges anyway, acting on many different strains, and strains of cholera bacteria are genetically very similar across the globe, increasing the chances that a cocktail of phages could be developed for these diseases that work across many patients and potentially even across entire continents. The specificity of phages, which target just the disease-causing species, while preserving other gut flora, is also an especially helpful characteristic when treating malnourished or immunocompromised populations.[4]

If the billions of dollars that is now pouring into fighting the antimicrobial resistance crisis can be directed to where it is truly needed – and an appreciation of the power of phages continues to take root in Africa – it is not far-fetched to imagine that in ten or twenty years, local scientists in phage clinics from Dakar to Harare will be up and away, brewing up their own unique and powerful antimicrobials from the rivers, sewers and the soils around them.

Perhaps, as the horrors of a post-antibiotic future start to bleed into the present, some countries will decide there is no time to put every phage therapy through years of clinical trials – leaning instead on the decades of experience of the Russians, Poles and Georgians. Rather than the high-tech vision of phage therapy being dreamed up in Europe and North America, the future of phage therapy in many parts of the world could be more 'DIY', with small clinics or roving microbiologists – perhaps even patients themselves – using the basic techniques pioneered by d'Herelle and Eliava to find low-cost treatments for bacterial infections in their own backyards.

The most promising way to use phages in the fight against drug-resistant bacteria may even be one that doesn't use phages at all. A number of research groups are exploring whether it is possible to take the viruses' most destructive antibacterial molecular weapons, concentrate them into a liquid and discard the rest. Lysins, for example, are the enzymes that phages use to pop open their bacterial hosts (or more accurately, the enzymes that their bacterial host is forced to make to pop itself). Several groups are investigating whether these compounds, used in isolation, could be used to kill bacteria without the need for the viruses themselves. Just like the phages they come from, these compounds can be highly specific, meaning different lysins could be used to kill specific species of bacteria, while leaving the body's 'good bacteria' unharmed. And just like phages, there are many different types of lysin out in the world waiting to be found when resistance to one type develops. With no self-replicating and infectious entity going anywhere near the patient, these biological products could also be produced and regulated more like traditional drugs.

Other groups are hoping to exploit the remarkable way that phages' tails are able to attach to and bore a hole in the tough outer coating of bacterial cells. These remarkable nano-scale structures, formed from tessellating proteins and synchronised chemical changes that turn groups of molecules into moving parts, are effectively microbial syringes, capable of puncturing and injecting material into cells. Scientists have already successfully created solutions of pure phage tails, minus their viral heads, to help explore the idea of using them as nanoscopic syringes to inject poison into bacterial cells. The idea clearly has potential, as it has been found that bacteria themselves also do exactly the same: some species have co-opted phage tails as a way to inject toxins into nearby, competing bacteria.[5]

In all likelihood, it's unlikely that any one of these approaches will be the magic bullet that replaces antibiotics and saves us from a return to the dark ages. Different infections in different healthcare systems in different parts of the world will require a different balance of high-tech and low-cost solutions. Difficult and unglamorous work to develop collaborations between phage biologists, clinicians, regulators and healthcare commissioners, and to refine and standardise the journey from diagnosis to treatment, will be just as important to getting whatever version of phage therapy is available off the ground in different parts of the world.

Besides, the judicious use of phages may actually help prolong the life of antibiotics. Studies suggest that a combination of phage therapy and antibiotics is more effective than either one on its own.[6] Phage therapy often causes drug-resistant bacteria to lose some of their resistance to antibiotics, as the microbes focus their resources on repelling the phages.[7] This means that even if the phage therapy is not completely effective, once-useless antibiotics are suddenly back in play in terms of treatment options. Some groups are even working on phages engineered to display molecules of antibiotic drugs on their heads, combining nature's deadly bacterial assassins with humanity's best chemical weapons, all in one go.

Even if we can make phage-based medicines that are proven to work and are economically viable and deliverable at scale, you also need patients to agree to take them. If, and when, phage therapy comes to a hospital or pharmacy near you, how will people react?

Having spoken to dozens of people who have been treated with phages – from super-informed scientist-patients who research and organise their treatment, to the shoulder-shrugging French truck driver Mr B – it's clear most patients' main concern is almost always whether it will work or not. But the roll-out of phage therapies or prophylactics on a massive scale could bring greater challenges.

In many regions, the word 'virus' remains synonymous with HIV and AIDS, and the world is not long out of a punishing viral pandemic. As demonstrated by COVID conspiracies, or the decades-long controversy about the measles, mumps and rubella (MMR) vaccine, misinformation and scare stories about viruses and medicine spread quickly and can be extremely difficult to counter. That's before we think of the influence of book, TV and film tropes involving apocalyptic disease outbreaks, mad scientists and zombies, which often have an escaped virus at their heart. The nightmare is a broad conspiracy about shadowy forces deploying strange viruses on the population.

It is not just patients who will need to be convinced, however: the biggest sceptics of phage therapy in the past have often been other scientists and doctors. From Felix d'Herelle to Carl Merril, advocates of phage therapy have time and time again faced prejudice and derision from their own colleagues. Pranav and Apurva Johri, whose Indian company Vitalis now helps facilitate access to treatment in Georgia, report doctors having 'aggressive reactions' or simply walking out of their presentations on phage therapy as recently as 2019. Members of the African Phage Forum reported senior professors dismissing their hopes to develop phage science in their institutions as 'a stupid idea' as recently as 2020. Virtually all the phage therapy patients I have interviewed told me that the doctors treating them originally had never even heard of phage therapy, let alone understood how they might access or use it – despite there being hundreds of papers on the topic just a few clicks away on Google. Phage therapy has a long, strange, and controversial history, and anyone writing about it – me included – seems unable to resist mentioning its dark and intriguing past. And so those negative associations with Stalin, science behind the Iron Curtain and economic hardship continue, regardless of efforts to modernise it.

In the opposite vein, it's also worth considering the danger of hype. Still, no major clinical trials have shown that phage therapy can be effective across a large population of patients. Yet as the buzz around phages grows, already we are seeing probiotics, cosmetic products and dietary supplements that claim to contain phages but don't, or seem to contain phages for no other reason than it sounds good.* It's worth remembering that part of the reason enthusiasm for phage therapy fizzled out in the 1930s and 1940s was the mass-marketing of phage products that didn't do half of the things they claimed they could.

Luckily, just about every misconception or misunderstanding there could be about phages has already played out over the last hundred years. Scientists like Professor Martha Clokie, a virologist and phage expert from the University of Leicester, have already started to think about how society might react to their little white pills being replaced with vials of viruses. She says that the various social and human factors that have hampered phage therapy throughout its history must be 'fully explored and understood' if we are to successfully advance the idea over the next hundred years – and has advised a period of 'pre-emptive reflection' on how the scientific community can inform and prepare the public for the arrival of this brave new phage-based world. Jeremy Barr, a phage expert from Monash

---

* On Amazon, you can now find various supplements that claim to contain 'life extension' phages or 'phage technology', with little to no explanation of what this actually means. And many in the phage community expressed concern when an extremely dubious-sounding medical technique – involving 'activating' the body's natural phages using 'electromagnetic frequencies' to cure Lyme disease – was presented at a respected conference in 2022.

University, Australia, has written about the need for a concerted global education programme before the widespread introduction of phage-based treatments – delivered through 'museum exhibitions, narrative literature, filmmaking and music' if necessary – to improve the public perception of phage therapy and provide clear and honest information about its promise and limitations.

It's impossible to know exactly how our relationship to phages will change over the next hundred years. Maybe phages will become known to the next generation as the technology and medicine that helps us control and manipulate the microbial world. Maybe things won't go so well, and in a post-antibiotic hellscape those of us who are left will still be talking about how the viruses of bacteria could save us, if only we could get a few more clinical trials up and running.

Perhaps phages will once again fall out of fashion, as new classes of chemical antibiotics rescue us from the full brunt of the resistance crisis just in time, or because other antibacterial strategies are developed that make phage therapy once again seem like more hassle than it is worth. Or maybe the use of phages in medicine, agriculture and in disinfectants becomes so widespread and wanton that we simply create another resistance problem all over again.

More likely, phages will become a much-needed part of an expanded toolkit against bacterial threats, used alongside existing antibiotics, new antibiotics and new methods to prevent infections in the first place – and hopefully they will be deployed in a way that is calculated to maximise effect while minimising or nullifying the inevitable new forms of resistance that arise.

What's clear is that many are now betting on a renaissance of phage science and technology in the coming years. In 2022, construction began on what will be the world's largest phage production plant, which will produce industrial quantities of phages not seen since the Eliava Institute in its heyday. The sleek, low, steel and glass factory rises out of the beautiful Lofoten Islands in northern Norway, and the company behind it, ACD Pharma, clearly believes phages will be needed in vast quantities in this part of Scandinavia soon. Initially, this is likely to be for the region's many fish farms, as they switch to using phages to reduce their reliance on antibiotics.* But eventually, phages could be produced here for many other uses, including phage-based medicines. But looking far into the future, it's not just infections in fish and people that phages could help us with. There's a whole world of other uses for these miraculous mini machines just waiting to be unleashed.

---

* Aquaculture (fish farming) is one of the fastest growing industries in the world, and antibiotics are often indiscriminately added to water or feed to prevent or treat disease among the fish stocks. The antibiotic-resistant bacteria that emerge soon transfer their resistance genes to bacteria of terrestrial animals and humans. Norway is seen as a leader in its efforts to regulate and reduce antibiotic use in fish farms – and phages are increasingly seen as the most promising alternative to these chemicals.

# 17

## The grey goo

In 2003, the heir to the British throne, Prince Charles (now King Charles III), asked the oldest scientific organisation in the world, the Royal Society, to investigate the risks of so-called 'nanotechnology' – that is, the engineering of materials and machines so small that they are measured in nanometres, or billionths of a metre. The ultimate aim of this field is to make devices so small that they can travel through human blood vessels, fighting disease, or integrate seamlessly into materials or the environment, doing useful things at a fundamental but largely unnoticeable scale.

The then-Prince of Wales, known for his support for environmental causes, was particularly worried about the dangers of so-called 'grey goo': a nightmare scenario where invisible, out-of-control, self-replicating nanobots proliferate so voraciously that they take over the entire planet. The source of his fears was a book by Eric Drexler, the visionary American engineer generally credited with popularising the term 'nano-technology' in the first place. Drexler had warned what would happen, in theory, if a nanomachine 1,000 times thinner than a human hair could copy itself in 1,000 seconds (roughly fifteen minutes). In ten hours, Drexler calculated, there would be 68 billion of them. 'In less than a day,' he warned, 'they would weigh a tonne, and in less than two days they would outweigh the Earth, and in another four hours they would exceed the mass of the sun and all the planets combined.'[8]

Of course, to consume the planet and the entire solar system, Drexler's self-replicating nanobots would require an endless source of raw materials and energy, which is silly. But as we have seen, the idea of invisible, self-replicating nano-machines that can copy themselves indefinitely is not silly at all. Some phages can make not just one of themselves but hundreds of themselves in fifteen minutes. A few phages, under the right conditions, can become a flask of trillions overnight. Feasting on the super-abundant bacteria in our world, they have ended up everywhere, in, on and around us, in immense numbers. In other words, the army of invisible, endlessly proliferating nanobots are already here. The 'grey goo' already took over this planet aeons ago.

After decades of trying to manipulate tiny amounts of certain molecules into complex forms – or sometimes build-ing with individual atoms – many in the field of nanotech-nology have come to realise that far better nanomachines than they could ever assemble already exist. In phages, they have an abundant supply of self-assembling, self-replicating, modifiable, DNA-injecting, environment-sensing and bacte-ria-killing nanobots, perfected over four billion years of trial and error, made entirely of organic material and largely non-toxic to humans.

Forget nanodevices painstakingly carved from silicon and carbon nanotubes; we have nanoscopic biological technology all around us. Phages are packed with miraculous molecular machinery, from the nanosyringes that inject the viral genes into cells, to the protein-based 'nanopumps' that help package the genetic material into the phage in the first place – which can squash so much DNA into the phage's head that the pressure inside is estimated to be thirty times greater than a car tyre.[9] Phages could even be described as ancient nanotechnologists themselves, using bacteria as their workshops to fashion all manner of clever structures from the complex chemical soup of nutrients and energy found in cells. (One phage, known as 'pharaoh', builds a tiny but perfectly formed pyramid on the outside of its host, from which new viruses emerge.[10])

So the question is, as our ever-increasing understanding of phages converges with advances in nanotechnology, genetic engineering and synthetic biology – what else might we be able to do with nature's army of tiny, programmable nanomachines?

The blood-brain barrier is a sheath-like membrane that coats all the blood vessels in your brain. Like a line of bouncers preventing anyone but VIPs from entering an exclusive area, it ensures that any old substance in the blood cannot simply diffuse into the delicate and easily spooked neurons that control our minds and bodies. This remarkable barrier lets in all the nutrients the brain needs, but stops 98% of small molecules and almost all large molecules from passing. That, unfortunately, includes most drugs. The brain's strict entrance policy is a major challenge to the development of therapies for disorders and diseases of the brain.

Remarkably, several different research groups have used phages as tiny vehicles to help smuggle medicines past this notorious membrane and into the brain – in lab animals at least. In one study by a team of researchers working at institutes in Detroit and New York, spherical *E.coli* phages, known as MS2, were coated in proteins that help trick the brain's security into letting them through. The phages were able to pass through the blood-brain barrier and accumulated deep in the tissues of the brain.[11] The researchers say the phages could easily be filled with pharmaceutical compounds to help diagnose or treat disorders of the mid-brain.

In another example, the heads of *Salmonella* phages (with the equally memorable name P22) were turned into brain-barrier-jumping 'nanocapsules' filled with the painkiller ziconotide.[12] This super-powerful painkiller, derived from strange venomous marine organisms known as cone snails and 1,000 times more powerful than morphine, can currently only be administered to treat severe pain by a horrifying injection directly into the spinal canal or the space between the skull and brain.

Helping drugs cross the blood-brain barrier is just one of many remarkable uses for phages being explored within the exciting world of 'nanomedicine' – where tiny particles, capsules or devices are used to deliver therapies around the body in the bloodstream. While many research groups have tried to develop nanomedicines using fancy man-made materials such as carbon nanotubes and metal nanoparticles, bacteriophages are really the ideal ready-made nanomedicines, having been perfected by nature over billions of years to protect, transport and inject material into a very specific target – and having already been proven to be mostly non-toxic to human systems.[13]

Not unlike the miniaturised submarines journeying through a person's body in the cult 1960s film *Fantastic Voyage*, the idea is that phages can carry a payload of medicine through the bloodstream before docking to the specific cells in the body where treatment is required. The phage's normal DNA cargo, carrying the genes required to make more of itself, are replaced by anti-cancer drugs or complex gene-editing compounds to help fix genetic diseases; the receptors on the phage's binding sites are replaced with molecules that target particular cells of the body, such as stem cells or tumour cells, for instance. Voilà: you have a nanoscopic, self-guiding drug-delivery vehicle.

The approach means that the often-toxic compounds used to treat cancer are directed to exactly where they are needed rather than washing through the entire body. In one study by scientists at the University of Tel-Aviv, modified worm-shaped phages were filled with anti-cancer drugs and engineered to bind only to cancer cells. Using lab-based models of cancerous tissue, the researchers found that the concentration of the drug around the tumour cells was over 1,000-fold higher than it would be if the drug was simply released into the bloodstream without its worm-phage chaperones.[14]

Other types of virus, including modified human viruses, have been explored as delivery vehicles for drugs in the past. But the ease with which phages can be grown (trillions at a time) and genetically engineered, their larger 'cargo' space, and their inability to replicate in human cells makes them particularly useful. Researchers have even created hybrids of phages and mammalian viruses to help phages deliver their medical payload to locations *inside* individual cells of the body, such as inside the nucleus of cells, where our DNA resides.[15]

Phages' potential use as nanoscopic medical tools doesn't stop there. Phages can also be engineered to display certain proteins on their outer coat, effectively turning them into travelling advertising boards to attract the attention of our immune systems. Several vaccines have been developed based on phages 'decorated' with molecules that prime the immune system against pathogens (although the success of the far simpler mRNA technology, at the heart of many of the most widely used COVID-19 vaccines, suggests these might never be needed).

In a totally different area of medicine, phages' unique self-assembling properties are showing promise in efforts to regrow damaged human tissue, such as skin, cartilage, bone or cardiac and neural tissue. In several studies, long, thin filamentous phages were modified to grow alongside each other like tiny fibres, forming a 3D-protein scaffold on which the human cells could grow.[16,17] Again, the phages can be grown trillions at a time and are easily modified to display helpful molecules on their outer coat – in this case proteins that help drive and organise the growth of human cells into natural structures and tissues. One group at the Zhejiang University in China is even using phages injected deep into the brain to help the brain rebuild itself following damage by stroke.[18] Phage-based 'nanofibres', loaded with stem cells, were injected directly into cavities in the brains of rats caused by strokes; the fibres provided an ideal scaffold for the stem cells to grow into new neurons and repair the damaged tissue.

Many companies are now looking at interesting uses of phage-based nanodevices and nanomaterials beyond medicine. According to some estimates, 'unwanted' bacteria are a $550 billion global problem,[19] affecting many different aspects of society and the environment. That astronomical figure includes the $42 billion antibiotics market, but also the value of products and services to stave off bacteria that cause disease or rot in our crops and food, contaminate drinking water, cause bad breath or body odour, clog up pipes and machinery and affect our pets' health. Plus, there's the socio-economic cost of the release of methane, a powerful greenhouse gas, by many different types of bacteria, often in animals' guts. With worrying reports that bacteria are even becoming resistant to essential sanitisers and disinfectants like ethanol,[20] many research organisations and companies are starting to use phages as living antibacterial nanoparticles that can be deployed in high-tech ways.

In a flat, low, octagonal building in a science park in Glasgow, Scotland, Dr Jason Clark excitedly tells me about what he describes as 'the most "sci-fi" device I've ever worked with'. Like something a comic book-mad scientist might use, a blue lightning bolt of electrical plasma explodes into the air from a high-voltage electrode and the machine throbs with an eerie purple glow. The high-energy plasma generated is known as 'corona discharge' and is used to make the surface of almost any material temporarily chemically active. This technology allows manufacturers to bind super-fine layers of material to another material's surface – ink logos to plastic food packaging, for example, or a fine chrome finish to plastic car parts. Clark places all manner of materials into his amazing plasma lightning bolt machine, but he's not sticking ink or chrome to them. He's adding a layer of phages.

Clark is chief scientific officer at Fixed Phage, a company that hopes to use this technology to create virus-coated nanomaterials for a range of different uses. Immobilising the phages on a surface stabilises them, meaning they don't have to be kept at a constant temperature, and potentially increasing their antibacterial activity from hours to days. Fixed Phage believe the approach can be used to zap phages onto all sorts of surfaces that need to be free from bacteria, for example medical devices, prosthetics and bandages, as well as food packaging and perhaps even food itself. One of the company's closest products to market is a type of thin plastic packaging for bags of pre-prepared salad. At the nanoscale, the surface of the plastic is covered in phages, and those that happen to be stuck tail-out are primed to inject DNA into any bacterial cells that brush up against them. This kicks off an epidemic of viral replication and bacterial destruction in the bags. 'What's nice is that nothing has been added to the salad bags that isn't there in the salads already,' Clark tells me. 'We just brew them up, stabilise them, attach them onto the packing and then put them back in again.'

As well as phage products for food packaging, fish farms, agriculture and medical devices, there are now phage-based nanoparticles for antibacterial mouthwashes, deodorants, toothpastes and cosmetics in development too. As scientists and healthcare teams are increasingly trying to understand the microbial communities that develop in even the cleanest hospitals, and find ways to make them less dangerous, even phage-impregnated *building materials* are being investigated. While attempts to make human phage therapy work struggle along, we could soon find products spiked with modified or fixed phages all over our food, cosmetics, homes and hospitals.

But the possibilities of phage-based nanotechnology are arguably even wilder – beyond virus-impregnated materials and little particles to deliver drugs round your bloodstream. What if phages could become the key components of a new blend of biology and technology?

Imagine, for example, phages festooned with gold, silver and platinum particles, crystals, nanoscopic carbon tubes and nano-magnets, like demented viral Christmas trees. Or phages made of conductive materials that act as tiny electronic devices. This is now commonplace in the world of 'nanobiotechnology', where things are as weird as they are tiny.

Increasingly, phages are being used as a kind of chassis upon which to build ever more complex nanoscale devices. They have already been used to create nanoscopic 'sensors' that use magnets or tiny resonating crystals to signal the presence of minute quantities of dangerous bacteria in the environment, or to detect the hallmark molecular markers of cancer and other diseases. Phages have been used as tiny moulds to grow phage-like structures made from other materials, including metals. ('Nanowires', for example, have been made from long filamentous phages that are covered in copper before the virus is incinerated, leaving just the sparkly, hollow, phage-shaped outer coating, which can then be used in ultra-tiny electronic devices smaller than a human chromosome.) The remarkable work of Professor Yoon Sung Nam and his 'Nano-Bio Interface Laboratory' at the Korean Advanced Institute of Science and Technology has even created phage-based 'microbatteries', capable of storing minute amounts of charged particles, and a tiny and self-assembling device that can generate hydrogen gas from sunlight and water. These ideas may not yet be efficient enough to have been developed into commercialised products, but the low material cost and low environmental impact of phage-based technology means it is likely to be an increasingly attractive technology in the future.

Which brings us back to the nightmarish idea of grey goo, which the former Prince of Wales was so concerned about. Might we one day release modified phages or self-replicating, phage-based nano-devices out into the environment, to proliferate on a vast scale and form a kind of amorphous living technology? Could an army of phages be released to quell outbreaks of dangerous bacteria in a vast region, to remove drug-resistance genes from waterways and ecosystems or even reverse the environmental degradation we have caused?

Most efforts to fix the environment by manipulating living systems on an ecosystem scale – known as biogeoengineering – have in the past been deemed to be too risky, with scientists generally concluding that such projects could have powerful unforeseen consequences on habitats and species that are already under huge pressure. However, scientists have started to explore whether modified phages can be deployed into large areas of land to alter the environment in useful ways.

In 2022, the PhageLand project[21] was granted funding to explore whether phages can be used to remove drug-resistance genes from vast areas of wetlands in Spain and Moldova, the first ever project of its kind. The reed beds of certain wetlands become hotspots for antibacterial resistance because they are often used to filter wastewater and sewage, where both bacteria and antibiotic run-off accumulates, and the PhageLand project will test whether specially developed phages can help selectively kill bacteria with antibacterial resistance genes in these places. It may well be the first of many projects that aim to deploy specially made phages out into the world and, despite sounding alarming, could be particularly useful for reducing environmental drug-resistance in lower income countries, where more complex and costly treatment plants are difficult to set up.

As well as the PhageLand project, researchers from the University of Delhi, writing in the journal *Ecology and Evolution*, have suggested that specially programmed phages could represent a 'tectonic shift for ecorestoration', used for a variety of environmental fixes, including as manipulators of the soil microbiome so that certain soils are more resistant to drought or able to break down pollutants.[22] Phages are also being explored as a way of altering the bacteria in farm animals' guts to reduce global methane emissions. More speculative ideas include using phages to help boost the amount of carbon taken up by marine ecosystems, or to spread genes among marine bacteria that will help them digest plastic.

Studies have already shown that some genetically engineered phages can outcompete natural phages or infect their hosts with a 1,000-fold greater efficacy, suggesting that if released they would rapidly spread throughout the environment.[23] Could we one day make use of the omnipresent power of environmental viruses to help fix the great ecological crises of our times? Pollution, biodiversity loss, even climate change? Could phages form the heart of a far stranger and as-yet-unimaginable technology that lives and integrates with our bodies and the natural world?

For the moment, such projects remain both controversial and firmly theoretical. The idea we can 'technofix' our way out of the current environmental crises is seen by many as an optimistic fantasy that distracts our efforts away from developing truly sustainable ways to exist on the planet, and in fact large-scale biogeoengineering projects, such as those aiming to boost how much carbon ocean life absorbs, are subject to a UN moratorium, given concerns about unintended consequences.[24] But if we do, one day, need an army of self-replicating nanobots to drastically alter our bodies or the world we live in from the bottom up, the grey goo is out there and waiting.

# Epilogue

## A new view of life

As I approach the end of almost two years researching the weird and wonderful world of phage science, I receive a link to a news report on an innovative new way of purifying water, being developed at Australia's University of the Sunshine Coast. It involves bacteriophages, deployed in water treatment works to remove unwanted bacteria. This 'new' idea goes back to the discoveries made at the very start of the first chapter – to the 1890s, when the moustachioed Ernest Hanbury Hankin first observed a substance able to purify the water of the Ganges.

That it has taken us over one hundred and thirty years to harness this natural purifying resource in our waterways says much about the strange and twisting fate of this field of science over the intervening years. The rollercoaster that is the last century of phage science has shown how the personalities of scientists, and the times they live through, can have a huge impact on the acceptance and impact of their discoveries. As much as we like to think of science as an empirical and objective process, it is conducted and organised by human beings – and human beings still have all manner of quirks, blind spots, prejudices and biases. Scientists are misunderstood, the significance of scientific work missed, and the pursuit of scientific knowledge is blown off course by the great geopolitical winds of the time. The viruses that can harm us have taken our attention away from the viruses that can heal us or do neither. And our view of viruses as not really 'alive' has further hampered our exploration of the phage-world we live in.

Despite the often-stuttering progress described in this book, we are now undoubtedly entering a truly exciting time for phage science. After so many missteps, misunderstandings, mistakes and misfortune, in the coming years we may finally see phage therapy available as a mainstream treatment in clinics and hospitals outside of the former Soviet bloc, new platforms to speed up the discovery and classification of phages, new phage-based technologies improving our lives and environments in interesting ways and the multitudinous phages of our planet given the study they deserve.

This book began with the suggestion that phages may be among the most important yet also most overlooked life forms on the planet. There are, of course, many important organisms that we ignore or neglect, but I think phages are a special case. We are talking about the most numerous biological entity on our planet, the greatest pool of unknown and unstudied genetic information in biology, endless potential allies in our fight against disease and decay and a foundational layer of the biosphere that has driven the diversity of life on Earth.

My other hope for this book was to recalibrate people's perceptions of what viruses are. How remarkable that a structure that looks so worrying, so *alien*, is in fact a founding member of life on our planet – one that augments and interacts with the world and with our lives in an inconceivably complex and epic way every day. Look at the totality of what phages have done for us just in the last one hundred years – their contribution to the molecular biology used every day in basic research, medicine, forensics and genetic engineering; their place at the heart of the powerful technology that might one day eradicate genetic disease; their role as alternative antibacterials. Look also at their role in our immune system – and you find that the number of lives saved each year by these unknown and unloved viruses may actually start to outnumber those taken by the viruses we know and fear. Add to that the enormous part phages have played in evolution, in ecosystems, in sharing genetic innovation among other life forms, and it becomes clear the virus can be, and is, a *good* thing, helping sustain and drive all the beauty we see in the natural world today.

I hope that, as your understanding of phages has changed, your understanding of the world has changed too. Our society is supported by fragile healthcare systems, built on shifting cultural and scientific sands – not to mention the ever-changing microbes we must fight. The complex microscopic and submicroscopic communities in the ocean, in the soil – and even in the human body and living on other life forms – are proving to be more diverse than we could ever have thought. (Remember those 848 different types of phage discovered just on the leaves of *wheat plants*.) The beautiful colour pictures of life on this planet have emerged from the background static of the planet's $10^{31}$ viruses.

These vast unknowns ensure there are many more exciting discoveries to come. Or, as one research paper put it, playfully misquoting Shakespeare: '*All the world's a phage.*'[25]

The next frontiers of phage science are not just about understanding the many trillions of unknown phages in unexplored parts of the world, but also the ones closer to home. Our familiarity with the phages in our gut and infection-prone areas of the body like the lungs is still in its infancy,* and how phages interact with our immune system and individual cells has long been overlooked. Understanding how phages and their hosts interact in complex environments, rather than just in Petri dishes, will be crucial in refining efforts to use phages more precisely. The close study of how phages form plaques – that most primitive visual symbol of the presence of phage that the very earliest phage hunters saw – is now helping us better understand how phages cut through biofilms, another key front in our battle against hardy and resistant bacteria.

Even within the phages themselves there are many secrets to unlock. Scientists do not have clear information on what most phages' genes actually do, even in well-studied species. And there are tantalising glimpses emerging of an even smaller class of parasites to explore, such as 'virophages' – viruses that infect giant viruses.[26]

On a different scale, we cannot yet predict whether the trillions of viruses in our oceans will exacerbate or attenuate the magnitude of climate change, but we know that they are a key element of the great chemical and energy cycles of this planet, and that they must be better understood.

Progress in all these areas could lead to a new era where the fundamental power and abundance of phages can be harnessed for the benefit of many areas of our lives.

I end with a plea not to repeat the mistakes of the past, not to see these viruses of bacteria as obscure and irrelevant. Just like the virus that lurked in a pangolin or bat that changed our world in 2020, or the viruses that forced microbes to evolve

---

* CrAssPhage01, a hyper-abundant gut phage that is probably the most abundant microbe in the human body, was only identified in 2014.

the CRISPR–Cas9 molecular system, they are anything but. 'The enemy of my enemy is my friend' goes the ancient Sanskrit proverb – and as the search for ways to fight bacteria intensifies, it's time to fully embrace these extraordinary, ubiquitous, invisible allies.

With great timing, just as I am concluding my final draft of the book, I receive the email I have been waiting for: my second phage-hunting effort had been a success. After my first botched attempt to isolate a phage at Exeter University, the team there had given me a second chance, and they have found several phages from a series of samples I took from a stream beside a petting zoo in my local town. Two are active against *E.coli* and three infect *Pseudomonas aeruginosa*, one of the six 'ESKAPE' pathogens causing particularly worrying drug-resistant infections around the world right now. At least three have unique-enough DNA sequences to say they are new to science and, a few weeks later, I even have pictures. I give them various silly names but call my favourite 'Emmi22', after my daughter, who babbled encouragement from the bank as I drew the water up in syringes.

It feels great to have made a small contribution to humanity's collective understanding of the planet's unfathomably vast viriome, and to have helped catalogue a few of the phages in my small corner of England. Who knows? Maybe one day Emmi22, or one of my other phages, will help save someone's life, whether in their raw form or just by the addition of their genes to the mass of data being used to drive phage therapy 2.0.

So, I think, now, I have written pretty much everything I can about the viruses of bacteria and why they are so important, and I've even gone out into the world and found my own. Now it's your turn.

# A field guide to phages

This spotter's guide shows just a handful of the most important and interesting phages we know of from the millions and possibly billions of different types out in the world.* Given that these structures are well under a micrometre in size, it's unlikely that you'll ever actually spot any of them, unless you own an electron microscope. Still, I hope you find wonder in knowing that these strange and sometimes sinister-looking structures are all around you, all over you and even inside you right now, hijacking bacterial cells, hitchhiking in their genomes, or cloning themselves a hundred-fold in the time it takes you to read the next dozen or so pages.

* As this book was being published, the way virologists classify phages was changed, and many of the classical phage 'families' such as Podovirus, Siphovirus and Myovirus, featured in this guide, were suddenly abolished. However, these terms will likely continue to be used to informally describe phages of similar shape and size.

# Escherichia virus T4

*Size: Approx. 200 nanometres long*
*Genome: Approx. 300 genes*
*Type: Myovirus*
*Host: E.coli*

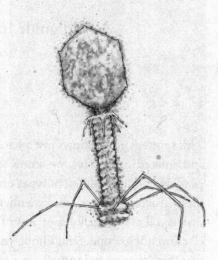

The celebrity of the bacteriophage world, Escherichia virus
T4's iconic, sinister outline can
be seen on posters, T-shirts
and even tattoos and jewellery.
It has become a kind of visual
shorthand for all viruses, even
human viruses, when in fact it
cannot infect anything except
*E.coli* bacteria.

Belonging to a group of tall-
tailed phages known as
myoviruses, T4 packages its
relatively large DNA genome
into an icosahedral (20-faced) head, and uses its long spider-like legs
(known as tail fibres) to sense suitable hosts. As the phage binds to its
host and punctures its outer wall, T4's rigid tail contracts to help inject
the DNA into its host like a syringe. It is one of the most well-studied
life forms ever, its structure understood almost down to the level of single
atoms.

# HTVC010P

*Size: Approx. 50 nanometres across*
*Genome: Approx. 60 genes*
*Type: Podovirus*
*Host: Pelagibacter ubique*

HTVC010P is probably the most abundant biological entity on the planet. Yes, it needs a better name.

Discovered only in 2013, this phage has evolved to infect the world's most dominant marine bacteria, *Pelagibacter ubique*. This bacteria is found in such huge numbers in the ocean that scientists once thought it must be immune to attack by phages. It is not. Every day, an unfathomably massive number of these bacteria are hijacked by HTVC010P and start doing little else but producing more HTVC010P.

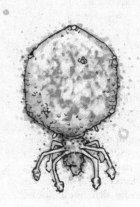

At certain times of the year, *Pelagibacter ubique* makes up as much as half of all the microbial cells in the ocean. These photosynthetic bacteria produce vast amounts of the oxygen that sustains life in the ocean and in the atmosphere. Phages that infect this bacteria, therefore, can have an impact on the availability of such vital nutrients on a planetary scale. The phage is an otherwise rather unremarkable member of the *Podoviridae* family of viruses which have extremely short, non-contracting tails, giving them a stubby, tick-like appearance.

# CTXφ

*Size: 1000+ nanometres long*
*Genome: Approx. 15 genes*
*Type: Filamentous*
*Host: Vibrio cholerae*

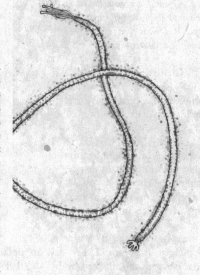

Another catchily named phage, CTXφ is a filamentous (thread-like) phage that is responsible for the deaths of millions of humans throughout history. The phage carries several genes that have super-charged the once harmless water-borne bacteria *Vibrio cholerae* into a much more deadly strain that causes the deadly diarrhoeal disease cholera.

CTXφ is a temperate phage, meaning it can lurk quietly inside its host rather than replicating madly and bursting it open. This means *Vibrio cholerae* infected with CTXφ continue to live and replicate while producing deadly toxins from CTXφ's genes as if they were its own.

CTXφ has taught us an important lesson that not all phages are 'good': they can often provide their host bacteria with genes that make them even more dangerous. Cholera kills up to 100,000 people every year, and it's all this skinny little phage's fault.

# ΦX174

*Size: 30 nanometres across*
*Genome: 11 genes*
*Type: Microvirus*
*Host: E.coli*

With just 11 genes, ΦX174 is one of the simplest life forms on Earth. Its genes contain the chemical recipe for 11 different proteins that together form a simple, self-replicating parasite capable of hijacking and reprogramming a bacterial cell.

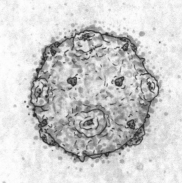

The virus particle is little more than a circular strand of DNA surrounded by 60 copies of the same protein molecule, tessellated into a protective shell, with 12 protein spikes to help it bind and inject its DNA into a host. It is thought that the phage's DNA is ejected through the middle of the spikes when the virus infects an *Escherichia coli* cell. Other proteins act as scaffolds to help the viral proteins assemble into the right shape.

ΦX174 exists at the fascinating boundary where complex chemistry becomes biology, where a certain arrangement of non-living molecules becomes capable of replication and evolution and therefore joins the ranks of what we call 'life'. Due to its simplicity, the first ever complete genome to be sequenced (that is, the unique sequence of subunits making up its DNA were read out) by molecular biologists was ΦX174. The first-ever completely 'synthetic organism', made using DNA synthesised chemically in a lab, was based on the super-minimal genome of ΦX174.

# Bicaudaviridae

*Size: Tails up to 400 nanometres in length*
*Genome: Approx. 70 genes*
*Type: Undefined*
*Host: Acidianus (archaea)*

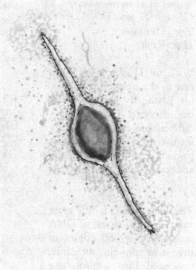

Also known as a 'two-tailed virus', Bicaudaviridae is one of many very weird phages that infect single-celled microbes known as archaea. Archaea, despite looking and behaving like bacteria, are in fact a quite distinct lineage of life that split from the common ancestor of all microbes billions of years ago. The viruses of archaea are lumped together with bacteriophages for convenience, but they are actually even less well studied or understood, and those that have been discovered have been found to be even stranger than bacterial viruses, often forming spindle, lemon or bottle-shaped viruses. They are often found in extreme environments of high salt, heat or pH.

Bicaudaviridae has evolved to be able to survive in extremely hot and acidic environments, such as hot springs, hydrothermal vents, mudpots or volcanic craters, where it infects the 'extremophile' archaea that have evolved to thrive in such places. Uniquely, when new viruses emerge from infected cells, they start out lemon-shaped and their tails then grow longer after they have been released into the world.

# Enterobacteria phage λ

*Size: capsid 50–60 nanometres across*
*Genome: Approx. 80 genes*
*Type: Siphovirus*
*Host: Escherichia coli*

Like the artist formerly known as Prince, this phage has become so important it is known just by a symbol, λ, known as lambda, the eleventh letter of the Greek alphabet. Since its discovery by Esther Lederberg in 1950, it has became one of the most useful organisms in molecular biology – used to study the fundamental processes that cells use to read and action the information contained in genes, and to help create genetically engineered organisms.

A phage that infects the microbiology lab workhorse *E.coli*, lambda helps researchers transport DNA into bacterial cells and integrate them into the bacterium's genome. In early genetic engineering, the virus was used to smuggle interesting stretches of DNA into *E.coli* cells to give the bacteria useful new functions, and it has been particularly useful in the study of how and why phages switch between the inconspicuous 'lurking' mode and the more violent cell-popping mode.

According to the journal *Virology*, λ's beauty is that it is simple enough to be 'potentially completely understandable . . . yet complex enough to be interesting and informative on many fronts.'

# Jumbo phages

*Size: Up to 600 nanometres long*
*Genome: Various*
*Type: Various*
*Host: Various*

In recent years, biologists have discovered some phages that blur the line between viruses and more complex cellular life. Most viruses are little more than a set of genetic instructions encased in a fancy protein vehicle and require the complexity of a host cell to replicate. But so-called jumbo phages appear different – much larger in size than most phages and far more complex.

These unusually large phages can have genomes up to 15 times longer than a typical

phage, which is larger than some bacteria – helping them to produce some of the more complex biochemical systems once thought to be found only in proper living cells. For example, jumbo phages can form a nucleus-like protective structure round their DNA and use bacterial defence enzymes to attack other viruses.

One of the largest phages ever found, Bacillus megaterium phage G, is around three times bigger than T4 (itself quite a large phage) and infects one of the largest known bacterial cells, *Bacillus megaterium*.

The discovery of hundreds of different jumbo phages (and, more recently, 'megaphages', which have even larger genomes) in many different environments has changed our understanding of what phages can be and the evolutionary relationship between viruses and cells. Are jumbo phages the missing link that show that viruses evolved *into* cells, that viruses evolved *from* cells, or that both viruses and cells have evolved from some ancient mix of the two?

# Acknowledgements

Ironically much of this book was written during the COVID-19 pandemic, when for long periods, travel and access to libraries and archives was impossible. I am therefore grateful for the excellent archive-digging and interviews conducted by William Summers, Alan Dublanchet, Ernst Fischer, Emiliano Fruciano, Anna Kuchment, Thomas Häusler, Dmitriy Myelnikov and Nina Chanishvili. Their works on the history of phage science are excellent and can be found in the notes.

I am also indebted to Dr Jess Sacher and Dr Sabrina Green for organising so many informal and open online meetings for the phage community, and for all their tweets about interesting happenings in phage science. These were an invaluable source of contacts and leads when face-to-face meetings were a no-go.

Thank you to Professor Martha Clokie and all who reviewed parts of my manuscript; Dr Ben Temperton and Dr Julie Fletcher at Exeter University for helping even silly old me find my very own phages; and Dr Antonia Sagona, Dr Tristan Ferry, Dr Zemphira Alavidze and Nina and Martha again for letting me spend time with you and your colleagues (and sometimes your patients). And thank you to all the phage scientists and patients who took time to chat to me over the past two years. Even if you were not directly quoted in the book, your expertise was useful and helped my understanding.

Huge thanks go to my agent Peter Tallack at Curious Minds (formerly the Science Factory) for helping develop a half-baked idea into the proposal that landed me my first book deal and to the Royal Society of Literature, who provided funding to help me complete the book through their Giles St Aubyn Prize for first-time writers. Thank you to my editors Anna Baty and Izzy Everington at Hodder, and Jessica Yao at Norton, not just for your astute wordcraft but for your energising enthusiasm for this project. Thank you also to Aruna Vasudevan for her excellent copy-edit.

Thank you to Dr Laura Bellingan and Dr Mark Downs at the Royal Society of Biology, who have both always supported me and my writing commitments beyond *The Biologist* magazine. I should also thank Jonn Elledge, without whom I might still be writing about the intricacies of doctors' contracts.

Most importantly, thank you to my very lovely wife, Jackie, for her unwavering support, understanding, help, ideas and love – much of it while we learned how to be parents during a pandemic. I'll stop talking about phages now, I promise.

And thank you to my parents, Steve and Dorrie, for the nearly four decades you have spent cheering me on from the sidelines, both metaphorically and literally.

# References & Notes

It has been difficult to know when to stop writing this book. Each week seems to bring new headlines about exciting new experimental phage therapy cases and new blockbuster grants for fundamental phage research, and there are important clinical trials of phage therapy due to report their results in the coming years. There are also countless interesting case studies and new ideas about phages that I could not squeeze into this book without it becoming unreadably vast.

For those who wish to find out more information on the development or rollout of phage therapy in different nations, the website bacteriophage.news provides useful updates and news articles, and a large database of private and publicly funded clinical studies can be found at clinicaltrials.gov (or the equivalent website in your country). The network phage.directory can help provide more information about accessing phage therapy for patients who have a referral or supporting documents from their doctor.

For an academic text covering many aspects of phage science, *Bacteriophages: Biology, Technology, Therapy*, edited by David Harper, Stephen Abedon and Benjamin Burrowes, (Springer, 2021) is the most comprehensive contemporary text on all things bacteriophage.

## Introduction and Part 1

1 Koonin, E. V., Martin W. 'On the Origin of Genomes and Cells within Inorganic Compartments.' *Trends in Genetics* 21(12), 647–54 (2005).

2 Yong, E. *I Contain Multitudes: The Microbes Within Us and a Grander View of Life*. HarperCollins 2016.

3 Reche, I., D'Orta, G., Mladenov, N. *et al.* 'Deposition Rates of Viruses and Bacteria above the Atmospheric Boundary Layer.' *The ISME Journal* 12, 1154–62 (2018).

4 'Phages Attack: A History of Bacteriophage Production and Therapeutic Use in Russia.' *Science First Hand*, May 2017, citing: Pokrovskaya M. P., *et al.* 'Lechenie ran bakteriofagom (Treatment of Wounds with Bacteriophage).' Moscow: USSR People's Commissariat of Public Health (Narkomzdrav), Medgiz (1941).

5 Bread was apparently only given out to the residents of Stalingrad after their 'phageing'. Chanishvili, N. & Alavidze, Z. Early Therapeutic and Prophylactic Uses of Bacteriophages. In Harper, D. R, Abedon, S. T., Burrowes B. H., & McConville, M. L. (eds) *Bacteriophages: Biology, Technology, Therapy* (Springer Reference, 2021).

6 Suttle C. A. 'Viruses in the Sea.' *Nature* 437(7057), 356-61 (2005).

7 Murray C., Ikuta K. S., Sharara F. *et al.* 'Global Burden of Bacterial Antimicrobial Resistance in 2019: A Systematic Analysis.' *The Lancet* 399(10325), 629–55 (2022).

8 'The Top 10 Causes of Death.' World Health Organization. 9 December 2020. www.who.int/news-room/fact-sheets/detail/the-top-10-causes-of-death

9 'The Ganges Brims with Dangerous Bacteria.' *The New York Times*. 23 December 2019.

10 Hankin M. E. 'L'action bactéricide des eaux de la Jumna et du Gange sur le vibrion du choléra.' *Annals of the Institut Pasteur (Paris)* 10, 511–23 (1896).

11 Faruque, S. M., Islam, M. J., Ahmad, Q. S. *et al.* 'Self-limiting Nature of Seasonal Cholera Epidemics: Role of Host-Mediated

Amplification of Phage.' *Proceedings of the National Academy of Sciences USA* 102(17), 6119–24 (2005).

12  Khairnar, K. 'Ganges: Special at its Origin.' *Journal of Biological Research (Thessaloniki)* 23, 16. (2016).

13  Kochhar, R. 'The Virus in the Rivers: Histories and Antibiotic Afterlives of the Bacteriophage at the Sangam in Allahabad.' *Notes & Records – The Royal Society Journal of the History of Science* 74(4), 625-651 (2020).

14  Abedon, S. T., Thomas-Abedon, C., Thomas, A. *et al.* 'Bacteriophage Prehistory: Is or Is Not Hankin, 1896, a Phage Reference?' *Bacteriophage* 1(3), 174–8 (2011).

15  Gamaleya, N. F. 'Bacteriolysins – Ferments Destroying Bacteria.' *Russian Archives of Pathology Clinical Medicine and Bacteriology* 6, 607–13 (1898).

16  From Nathan Bailey's Dictionary 1770, with various other definitions sourced from the Online Etymology Dictionary, etymonline.com.

17  Twort, A. *In Focus, Out of Step: A Biography of Frederick William Twort F.R.S. 1877–1950* (Alan Sutton Publishing, 1993).

18  Duckworth, D., 'Who Discovered Bacteriophage?' *Bacteriological Reviews* (1976), citing Twort, F. W. 'The Discovery of the Bacteriophage.' *Penguin Sci. News* 14, 33–4 (1949).

19  Summers, W. C. 'Félix Hubert d'Herelle (1873–1949): History of a Scientific Mind.' *Bacteriophage* 6(4), (2016).

20  Felix d'Herelle's unpublished memoirs, *Les pérégrinations d'un microbiologiste* ('the wanderings of a microbiologist') were written between 1940 and 1946 and are currently held in the Pasteur Institute Archives, Paris. Much of the passages quoted here are translations found in Summers, W. C. *Félix d'Herelle and the Origins of Molecular Biology* (New Haven and London: Yale University Press, 1999), or Summers (2016), above.

21  Summers, W. C. *Félix d'Herelle and the Origins of Molecular Biology* (New Haven and London: Yale University Press, 1999).

22  See Summers (1999).

23  See Summers (1999).

24 Thomas, G. H. 'William Twort: Not Just Bacteriophage.' *Microbiologysociety.org*. Accessed 29 May 2014.

25 See Summers (2016).

26 Correspondence with Nina Chanishvili of the Eliava Institute of Bacteriophages, Tbilisi, January 2021.

27 See Summers (2016).

28 See Twort (1993).

29 Fruciano, E. Bacteriophage research: the causes and the effects of the conflict between Felix d'Herelle and the Pasteur Institute (1917–1949). [Citations 104 and 105]. This draft manuscript is based on a collection of unpublished documents and letters from d'Herelle and colleagues at the institute.

30 See Summers (1999).

31 From the assorted d'Herelle family diaries, cited in Thomas Häusler, *Viruses vs Superbugs* (London: Macmillan, 2006).

32 Twort, F. W. 'An Investigation on the Nature of Ultra-Microscopic Viruses.' *The Lancet* 186(4814), 1241–3 (1915).

33 See Thomas (2014).

34 See Twort (1915).

35 'Find Way to Rid World of Locusts; French Doctor's Campaign of Extermination in Argentina a Complete Success.' *The New York Times*, 11 July 1912.

36 See Summers (1999).

37 See Summers (2016).

38 Cruz, B. *et al.* 'Quantitative Study of the Chiral Organization of the Phage Genome Induced by the Packaging Motor.' *Biophysical Journal* 118, 2103–16 (2020).

39 Dublanchet, A. 'The Epic of Phage Therapy.' *Canadian Journal of Infectious Diseases and Medical Microbiology* 18(1), 15–18 (2007).

40 D'Herelle, F. 'On an Invisible Microbe Antagonistic to Dysentery Bacilli.' Note by Mr F. d'Herelle presented by M. Roux. *Comptes rendus Academie des Sciences* 165, 373–5 (1917)

41 Dublanchet, A. 'The Epic of Phage Therapy.' *Canadian Journal of Infectious Diseases and Medical Microbiology* 18(1), 15–18 (2007).

42 Fruciano, E. (unpublished).

43 Häusler, T. *Viruses vs Superbugs: A Solution to the Antibiotics Crisis?* (Macmillan, 2006), 96.

44 See Dublanchet (2007).

45 Fleming named the substance 'lysozyme'. It interferes with the carbohydrates in cell walls and only has a mild antiseptic effect. Fleming, A. & Allison, V. D. 'Observations on a Bacteriolytic Substance ("Lysozyme") Found in Secretions and Tissues.' *British Journal of Experimental Pathology* 3(5), 252–60.

46 From d'Herelle's unpublished memoirs, in Summers (1999).

47 See Häusler (2006), 68.

48 See Häusler (2006), 49.

49 See Häusler (2006), 49.

50 Bruynoghe R., Maisin J. 'Essais de thérapeutique au moyen du bacteriophage.' *C R Soc Biol.* 85, 1120–1 (1921). The paper citing it as the first is Sulakvelidze, A. *et al.* 'Bacteriophage Therapy.' *Antimicrobial Agents and Chemotherapy* 45(3), 649–59 (2001).

51 See Fruciano, E. (2007).

52 See Fruciano, E. (2007).

53 From the assorted d'Herelle family diaries, cited in Häusler (2006).

54 See Fruciano, E. (2007).

55 See Summers (1999).

56 See Fruciano, E. (2007).

57 See Fruciano, E. (2007).

58 See Summers (1999).

59 See Fruciano, E. (2007).

60 Numbers are based on the 1934 review by Eaton & Bayne-Jones. 'Bacteriophage Therapy: Review of the Principles and Results of the Use of Bacteriophage in the Treatment of Infections.' *JAMA* 23, 1769–76 (1934).

61 See Fruciano, E. (2007).

62 Letarov, A. V. 'History of Early Bacteriophage Research and Emergence of Key Concepts in Virology.' *Biochemistry Moscow* 85 (9), 1093–1112 (2020).

63 See Fruciano, E. (2007).

64 Fruciano, E. (unpublished) – letters between Gracia and Calmette from 1930, notes 136–7.

65 Fruciano, E. (unpublished) and www.nobel.se.

66 Dublanchet A., Fruciano, E. 'Félix d'Herelle et les pasteuriens.' *Association des anciens élèves de l'Institut Pasteur* 193, 170–4 (2007).

67 Chanishvili, N. & Goderdzishvili, M. 'Commercial Products for Human Phage Therapy.' Chapter 5 in Coffey, A. & Buttimer, C. (eds). *Bacterial Viruses: Exploitation for Biocontrol and Therapeutics* (Caister Academic Press, 2019).

68 See Summers (1999).

69 Report of Colonel Anderson to the Cholera Committee of the Indian Research Fund Association. In Häusler (2006), 103.

70 'The Use of Bacteriophage.' *Science* 70(1817), x–xii (1929).

71 See Häusler (2006), 92.

## Part 2

1 Chanishvili, N. & Goderdzishvili, M. 'Commercial Products for Human Phage Therapy.' Chapter 5 in Coffey, A. & Buttimer, C. (eds). *Bacterial Viruses: Exploitation for Biocontrol and Therapeutics* (Caister Academic Press, 2019).

2 Biographical details are from Chanishvili, N. 'Phage Therapy – History from Twort and d'Herelle through Soviet Experience to Current Approaches.' *Advances in Virus Research* 83, 3–40 (2012).

3 Kuchment, A. *The Forgotten Cure* (Springer, 2012), 51.

4 See Summers (1999), 152.

5 See Summers (1999), 153.

6 Myelnikov D. 'An Alternative Cure: The Adoption and Survival of Bacteriophage Therapy in the USSR, 1922–1955.' *Journal of the History of Medicine and Allied Sciences* 73(4), 385–411 (2018). Quoting: 'V SSSR vtorichno priezzhaet professor d'Errel.' *Pravda* 298, no.6184 (28 October 1934).

7 D'Herelle, F. *Le Phénomène de la Guérison dans les Maladies Infectieuses* (Masson et cie, Paris, 1938).

8 Translated quotation from Summers (1999).

9  Interview with Dr Nina Chanishvili, January 2021.

10 Shrayer D. P. 'Felix d'Herelle in Russia.' *Bulletin De L'Institut Pasteur* 94, 91–6 (1996).

11 Field, Mark G. *Soviet Socialized Medicine: An Introduction* (New York: Free Press; London: Collier-Macmillan, 1967) 51–2.

12 See Field (1967).

13 See Field (1967).

14 See Field (1967).

15 See Chanishvili, N. (2012).

16 Myelnikov, D. 'Creature Features: The Lively Narratives of Bacteriophages in Soviet Biology and Medicine.' *Notes and Records of the Royal Society of London* 74(4), 579–97 (2020).

17 Myelnikov, D. 'An Alternative Cure: The Adoption and Survival of Bacteriophage Therapy in the USSR, 1922–1955.' *Journal of the History of Medicine and Allied Sciences* 73(4), 385–411 (2018).

18 Interview with Nina Chanishvili at the Eliava Institute, Tbilisi, May 2022.

19 See Chanishvili (2012).

20 See Chanishvili (2012).

21 See Myelnikov (2018)

22 See Myelnikov (2018)

23 See Myelnikov (2018)

24 See Myelnikov (2018)

25 See Myelnikov (2020)

26 Krueger, A. P., and Scribner, E. J. 'Bacteriophage Therapy. II. The Bacteriophage: Its Nature and its Therapeutic Use.' *JAMA* 19, 2160–277 (1941).

27 Interview with Carl Merril, September 2020.

28 See Häusler (2006), 108.

29 See Häusler (2006), 109.

30 See Kuchment (2012), 98.

31 'Phages Attack: A History of Bacteriophage Production and Therapeutic Use in Russia.' *Science First Hand* 46(1), (2017) https://scfh.ru/en/papers/phages-attack/

32 See Myelnikov, 2018.

33  See Myelnikov, 2018.

34  See Myelnikov, 2018.

35  See Kuchment (2012), 99.

36  Chanishvili, N. & Alavidze, Z. 'Early Therapeutic and Prophylactic Uses of Bacteriophages. In Harper, D. R., Abedon, S. T., Burrowes, B. H., & McConville, M. L. (eds) *Bacteriophages: Biology, Technology, Therapy* (Springer Reference, 2021).

37  De Freitas Almeida, G. M., & Sundberg, L. 'The Forgotten Tale of Brazilian Phage Therapy.' *Historical Review* 20(5), 90–101 (2020).

38  Ry Young, as quoted in Strathdee, S. *The Perfect Predator: A Scientist's Race to Save Her Husband from a Deadly Superbug: A Memoir* (Hachette, 2019).

39  Ventola C. L. 'The Antibiotic Resistance Crisis: Part 1: Causes and Threats.' *Pharmacy and Therapeutics* 40(4), 277–83 (2015).

40  Sir Alexander Fleming – Nobel Lecture. NobelPrize.org. https://www.nobelprize.org/prizes/medicine/1945/fleming/lecture/

41  Baker, K. S., Mather, A. E., McGregor, H. *et al.* 'The Extant World War 1 Dysentery Bacillus NCTC1: A Genomic Analysis.' *The Lancet* 384(9955), 1691–7 (2014).

42  Paun, V. I., Lavin, P., Chifiriuc, M. C. *et al.* 'First Report on Antibiotic Resistance and Antimicrobial Activity of Bacterial Isolates from 13,000-Year-Old Cave Ice Core.' *Scientific Reports* 11(514), (2021).

43  See Kuchment (2012), 99.

44  Hoyle, N., Fish, R., Nakaidze, N. *et al.* 'An Overview of Current Phage Therapy: Challenges for Implementation.' Chapter 1 in Coffey, A. & Buttimer, C. (eds). *Bacterial Viruses: Exploitation for Biocontrol and Therapeutics* (Caister Academic Press, 2019).

## Part 3

1  See Häusler (2006), 176.

2  Parfitt, T. 'Georgia: An Unlikely Stronghold for Bacteriophage Therapy.' *The Lancet* 365(9478), 2166–7, (2005).

3  See Häusler (2006), 190.

4  Osborne L. 'A Stalinist Antibiotic Alternative'. *The New York Times Magazine*, 6 February 2000.

5  Betty Kutter interview by Martha Clokie in *PHAGE: Therapy, Applications and Research* 1(1), (2020).

6  See Kuchment (2012), 136

7  Radetsky, P. 'The Good Virus'. *Discover Magazine*. 1 November 1996. www.discovermagazine.com/technology/the-good-virus

8  See Kuchment (2012), 139

9  See Kuchment (2012), 139.

10  See Häusler (2006), 10.

11  'Our First Adventure With Phage Therapy: Alfred's Story'. (Betty Kutter, Evergreen Bacteriophage Lab blog – date unknown.) https://sites.evergreen.edu/phagelab/about/alfreds-story

12  See Häusler (2006), 8–9.

13  See Kuchment (2012), 5.

14  See Osborne (2000).

15  See Häusler (2006), 10.

16  See Häusler (2006), 12.

17  See Häusler (2006), 13.

18  Interview with Zemphira Alavidze, July 2021.

19  Parfitt, T., 'Georgia: An Unlikely Stronghold for Bacteriophage Therapy.' *The Lancet* 365(9478), 2166–7, (2005).

20  See Parfitt (2005).

21  *The Sopranos*, Series 6 Episode 19: 'The Second Coming.'

22  See Kuchment (2012), 136.

23  Chanishvili, N. *A Literature Review of the Practical Application of Bacteriophage Research* (Nova Biomedical, 2012).

24  Chanishvili, N. & Goderdzishvili, M. 'Commercial Products for Human Phage Therapy.' Chapter 5 in Coffey, A. & Buttimer, C. (eds). *Bacterial Viruses: Exploitation for Biocontrol and Therapeutics* (Caister Academic Press, 2019).

25  Interview with Dr Randy Fish by Jess Sacher on successfully treating diabetic foot wounds with phages. Phage Directory, 14 April 2021. www.youtube.com/watch?v=tOdoEVY0cq0

26 Fish, R., Kutter, E., Wheat, G., *et al.* 'Bacteriophage Treatment of Intransigent Diabetic Toe Ulcers: A Case Series.' *Journal of Wound Care* 25(7), S27–S33 (2016).

27 Merril, C. R., Geier, M. R. & Petricciani, J. C. 'Bacterial Virus Gene Expression in Human Cells.' *Nature* 233, 398–400, (1971); Merril, C. R., Geier, M. R. & Petricciani, J. C. 'Bacterial Gene Expression in Mammalian Cells.' In G. Raspe (ed.), *Advances in the Biosciences* 8, 329–42, (1972).

28 Merril, C. R. *et al.* 'Isolation of Bacteriophages from Commercial Sera.' *In Vitro* 8, 91–3, (1972).

29 Kolata, G. B. 'Phage in Live Virus Vaccines: Are They Harmful to People?' *Science* 187(4176) 522–3 (1975).

30 Merril's personal recollections and quotes in this chapter are from interviews between he and I in September 2020 and April 2022. Additional information was obtained from an interview with Biswajit Biswas of the Naval Medical Research Center, Frederick, in December 2021.

31 'F.D.A. Finds Four Vaccines Contaminated With Probably Harmless Viruses.' *The New York Times*, 4 May 1973.

32 'Food and Drug Administration Rules and Regulations.' Vol. 38, 1973. 11080e1 – 'Certain Viral Vaccines Containing Unavoidable Bacteriophage.' Federal Register.

33 Merril, C. R., Dunau, M. L. & Goldman, D. 'A Rapid Sensitive Silver Stain for Polypeptides in Polyacrylamide Gels.' *Anal. Biochem.* 110(1), 201–7, 1981.

34 Merril, C. R. *et al.* 'Long-Circulating Bacteriophage as Antibacterial Agents.' *Proceedings of the National Academy of Sciences USA* 93, 3188–92 (1996).

35 Merril, C. R., Scholl, D. & Adhya, S. L., 'The Prospect for Bacteriophage Therapy in Western Medicine.' *Nature Reviews: Drug Discovery* 2, 489–97 (2003).

36 'How the Navy brought a once-derided scientist out of retirement — and into the virus-selling business.' *STAT News*, 16 October, 2018.

37 The details of Tom Patterson's illness and treatment with phages are from Strathdee, S. *The Perfect Predator: A Scientist's Race to Save*

*Her Husband from a Deadly Superbug: A Memoir* (Hachette, 2019), plus various media coverage of the case and an interview with Steffanie Strathdee, January 2021.

38 Chan, B. K. *et al*, Phage treatment of an aortic graft infected with Pseudomonas aeruginosa. *Evolution, Medicine, and Public Health* 2018(1), 60–6 (2018). See also: 'A virus, fished out of a lake, may have saved a man's life — and advanced science.' *STAT News*, 7 December 2016.

39 Interview with Cara Fiore, Senior Regulatory Reviewer for the FDA, January 2022.

40 Interview with Steffanie Strathdee, UC San Diego, January 2021.

41 See Strathdee (2019), 170.

42 Miller H. 'Phage Therapy: Legacy of CF Advocate Mallory Smith Endures.' *Cystic Fibrosis News Today*, 27 May 2021.

43 Interview with Pranav and Apurva, November 2021, and www.vitalisphagetherapy.com

44 Loponte R., Pagnini U., Iovane G., *et al*. 'Phage Therapy in Veterinary Medicine.' *Antibiotics* 10(4), 421 (2021).

45 Yost, D. G., Tsourkas P., & Amy P. S. Experimental bacteriophage treatment of honeybees (*Apis mellifera*) infected with *Paenibacillus larvae*, the causative agent of American Foulbrood Disease. *Bacteriophage* 6(1), e1122698 (2016).

46 Greene, W., Chan, B., Bromage, E. *et al*. 'The Use of Bacteriophages and Immunological Monitoring for the Treatment of a Case of Chronic Septicemic Cutaneous Ulcerative Disease in a Loggerhead Sea Turtle *Caretta caretta*.' *Journal of Aquatic Animal Health* 33(3) 139-54 (2021).

47 Ferriol-González C., & Domingo-Calap P. 'Phage Therapy in Livestock and Companion Animals.' *Antibiotics* (Basel) 10(5), 559 (2021).

48 'University of Leicester research aims to "save silk trade in India".' BBC News Online, 10 January 2016.

49 Wright A., Hawkins C. H., Anggård E. E., *et al*. 'A Controlled Clinical Trial of a Therapeutic Bacteriophage Preparation in Chronic Otitis

Due to Antibiotic-resistant Pseudomonas Aruginosa; a Preliminary Report of Efficacy.' *Clinical Otolaryngology* 34(4), 349–57 (2009).

50 Rhoads, D. D., Wolcott, R. D., Kuskowski, M. A. *et al.* 'Bacteriophage Therapy of Venous Leg Ulcers in Humans: Results of a Phase I Safety Trial.' *Journal of Wound Care* 18(6), 237–8 (2009).

51 Sarker S. A., Sultana S., Reuteler G. *et al.* 'Oral Phage Therapy of Acute Bacterial Diarrhea with Two Coliphage Preparations: A Randomized Trial in Children from Bangladesh.' *EBioMedicine.* 4, 124-37 (2016).

52 Jault P., Leclerc T., Jennes S. *et al.* Efficacy and Tolerability of a Cocktail of Bacteriophages to Treat Burn Wounds Infected by Pseudomonas Aeruginosa (PhagoBurn): A Randomised, Controlled, Double-Blind Phase 1/2 Trial. *The Lancet Infectious Diseases* 19, 35–45 (2019).

53 Leitner, L., Ujmajuridze, A., Chanishvili, N. *et al.* 'Intravesical bacteriophages for treating urinary tract infections in patients undergoing transurethral resection of the prostate: a randomised, placebo-controlled, double-blind clinical trial.' *The Lancet Infectious Diseases* 21(3), 427–6 (2021).

54 Chanishvili, N., Nadareishvili, L., Zaldastanishvili, E. *et al.* 'Application of Bacteriophages in Human Therapy: Recent Advances at the George Eliava Institute.' Chapter 6 in Coffey, A. & Buttimer, C. (eds). *Bacterial Viruses: Exploitation for Biocontrol and Therapeutics* (Caister Academic Press, 2019).

55 Eskenazi, A., Lood, C., Wubbolts, J. *et al.* 'Combination of Pre-Adapted Bacteriophage Therapy and Antibiotics for Treatment of Fracture-related Infection Due to Pandrug-resistant Klebsiella pneumoniae.' *Nature Communications* 13, 302 (2022).

56 Little, J. S., Dedrick, R. M., Freeman, K. G. *et al.* 'Bacteriophage Treatment of Disseminated Cutaneous *Mycobacterium chelonae* Infection.' *Nature Communications* 13, 2313 (2022).

57 Dedrick R. M., Guerrero-Bustamante C. A., Garlena R. A. *et al.* 'Engineered Bacteriophages for Treatment of a Patient with a Disseminated drug-resistant *Mycobacterium abscessus.*' *Nature Medicine* 25(5), 730–3 (2019).

58 Sacher, J. C., Zheng J., & Lin R. C. Y. 'Data to Power Precision Phage Therapy: A Look at the Phage Directory–Phage Australia Partnership.' *PHAGE* 3(2), 112-15 (2022).


## Part 4

1 Cairns, J., Stent G. S., Watson J. D. (eds). *Phage and the Origins of Molecular Biology* (Cold Spring Harbor Laboratory, Cold Spring Harbor, New York, 1966), 57.

2 Cobb, M. *Life's Greatest Secret: The Race to Crack the Genetic Code* (Profile Books, 2015), 16.

3 Hayes, W. *Biographical Memoir of Max Delbrück* (Washington D.C. National Academy of Science, 1993).

4 Fischer, E. P. & Lipson, C. *Thinking About Science: Max Delbrück and the Origins of Molecular Biology* (New York, Norton, 1988). Family numbering, 17, Quote, 20.

5 Harding, C. 'Max Delbrück (1906–1981)' Interviewed by Carolyn Harding. *California Institute of Technology Archives and Special Collections* (1978).

6 See Fischer (1988), 25.

7 See Harding (1978), 29.

8 See Harding (1978), 30 & Fischer (1988), 50.

9 See Fischer (1988), 82–3.

10 Schrödinger, E. *What is Life? The Physical Aspect of the Living Cell* (Cambridge University Press, 1944).

11 See Harding (1978), 57–8.

12 See Harding (1978), 58.

13 See Harding (1978), 109.

14 See Harding (1978), 60.

15 See Harding (1978), 63.

16 See Harding (1978), 117.

17 See Fischer (1988), 114.

18 Luria, S. E. *A Slot Machine, a Broken Test Tube: An Autobiography* (HarperCollins, 1984).

19 See Fischer (1988), 146.

20 See Fischer (1988), 148.

21 See Fischer (1988), 150.

22 See Fischer (1988), 148.

23 Stahl, F. *Alfred Day Hershey: A Biographical Memoir* (The National Academy Press, 2001).

24 *Albert Hershey. The Max Delbrück Laboratory Dedication Ceremony* (New York: Cold Spring Harbor, 1981).

25 'Alfred D. Hershey, Nobel Laureate for DNA Work, Dies at 88.' *The New York Times,* 24 May 1997.

26 Ackermann H.W. 'The First Phage Electron Micrographs.' *Bacteriophage* 1(4), 225–7 (2011).

27 Ellis, E. & Delbrück, M. 'The Growth of Bacteriophage.' *Journal of General Physiology* 22(3), 365–84 (1939).

28 Luria, S. E., Anderson, T. F. 'The Identification and Characterization of Bacteriophages with the Electron Microscope.' *Proceedings of the National Academy of Sciences USA* 28, 127–30 (1942).

29 See Ackermann (2011).

30 See Fischer (1988), 134.

31 See Cairns (1966), 68.

32 Judson, H. F. *The Eighth Day of Creation* (Cold Spring Harbor Press, 1996).

33 See Cairns (1966), 302.

34 Fischer (1988), 158.

35 Fischer (1988), 179.

36 See Cairns (1966), 63.

37 Morange, M. 'What History Tells Us III. André Lwoff: From Protozoology to Molecular Definition of Viruses.' *J. Biosci.* 30, 591–4 (2005).

38 Lwoff, A. (1966) 'The Prophage and I.' in: *Phage and the Origins of Molecular Biology* (Cold Spring Harbor Laboratory Press, New York) 88–99.

39 Abedon, S. T. 'The Murky Origin of Snow White and her T-even Dwarfs.' *Genetics* 155(2), 481–6 (2000).

40 Interview with E. P. Fischer, November 2020.

41 Avery, O. T., MacLeod, C. M., *et al.* 'Studies on the Chemical Nature of the Substance Inducing Transformation of Pneumococcal Types: Induction of Transformation by a Deoxyribonucleic Acid Fraction Isolated from Pneumococcus Type III.' *Journal of Experimental Medicine* 79(2), 137–58 (1944).

42 Hershey, A., & Chase, M. 'Independent Functions of Viral Protein and Nucleic Acid in Growth of Bacteriophage.' *Journal of General Physiology* 36(1), 39–56 (1952).

43 See Judson (1996).

44 See Judson (1996), 40.

45 Watson, J. D. *The Double Helix* (Signet Books, 1969), 345.

46 See Fischer (1988), 198.

47 See Watson (1969), 1226.

48 See Fischer (1988), 200.

49 Based on my own analysis of information available at nobelprize. org, including prize information and Nobel laureate biographies & essays.

50 Roux, S., Brum, J., Dutilh, B. *et al.* 'Ecogenomics and Potential Biogeochemical Impacts of Globally Abundant Ocean Viruses.' *Nature* 537, 689–93 (2016).

51 Staley, J. T. & Konopka, A. 'Measurement of In Situ Activities of Nonphotosynthetic Microorganisms in Aquatic and Terrestrial Habitats.' *Annual Review of Microbiology* 39, 321–46 (1985).

52 Bergh, O., Børsheim, K. Y., Bratbak, G. *et al.* 'High Abundance of Viruses Found in Aquatic Environments.' *Nature* 340, 467–8 (1989).

53 Proctor, L. M. & Fuhrman, J. A. 'Viral Mortality of Marine Bacteria and Cyanobacteria.' *Nature* 343, 60–2 (1990).

54 Suttle, C. A., Chan, A. M., Cottrell, M. T. 'Infection of Phytoplankton by Viruses and Reduction of Primary Productivity.' *Nature* 347, 467–9 (1990).

55 Breitbart, M., Bonnain, C., Malki, K. *et al.* 'Phage Puppet Masters of the Marine Microbial Realm.' *Nature Microbiology* 3, 754–66 (2018).

56 See Suttle (2005).

57 Weinbauer, M. G. & Rassoulzadegan. F. 'Are Viruses Driving Microbial Diversification and Diversity?' *Environmental Microbiology.* 6(1), 1–11 (2004).

58 See Breitbart (2018).

59 See Breitbart (2018).

60 Warwick-Dugdale, J., Buchholz, H. H., Allen, M. J. *et al.* 'Host-hijacking and Planktonic Piracy: How Phages Command the Microbial High Seas.' *Virology Journal* 16, 15 (2019).

61 Zimmer, C. *Planet of Viruses* (University of Chicago Press, 2011), 59.

62 Dion, M. B., Oechslin, F. & Moineau, S. 'Phage Diversity, Genomics and Phylogeny.' *Nature Reviews Microbiology* 18, 125–38 (2020).

63 Dennehy J. J. & Abedon, S. T. 'Bacteriophage Ecology.' Chapter in Harper, D. R, Abedon, S. T, Burrowes B. H, & McConville, M. L. (eds) *Bacteriophages: Biology, Technology, Therapy* (Springer Reference, 2021), 268.

64 Reche, I. *et al.* 'Deposition Rates of Viruses and Bacteria above the Atmospheric Boundary Layer.' *The ISME Journal* 12(4), 1154–62 (2018).

65 Forero-Junco L. M., Alanin K. W. S., Djurhuus A. M. *et al.* 'Bacteriophages Roam the Wheat Phyllosphere.' *Viruses* 14(2), 244 (2022).

66 Działak, P., Syczewski, M. D., Kornaus, K. 'Do Bacterial Viruses Affect Framboid-like Mineral Formation?' *Biogeosciences*, 19 4533-50 (2022).

67 Suttle, C. 'Marine Viruses — Major Players in the Global Ecosystem.' *Nature Reviews Microbiology* 5, 801–12 (2007).

68 Breitbart M. 'Marine Viruses: Truth or Dare.' *Annual Review of Marine Science* 4, 425–48 (2012).

69 Chibani-Chennoufi, S., Bruttin, A., Dillmann, M. *et al.* 'Phage-host interaction: An Ecological Perspective.' *Journal of Bacteriology.* 186(12), 3677–86 (2004).

70 Groman, N. B. 'Conversion by Corynephages and Its Role in the Natural History of Diphtheria.' *Journal of Hygiene, Cambridge* 93, 405–17 (1984).

71 Rosenwasser, S., Ziv, C. *et al.* 'Virocell Metabolism: Metabolic Innovations During Host-Virus Interactions in the Ocean.' *Trends in Microbiology* 10, 821–32 (2016).

72 Zhao, Y., *et al.* 'Abundant SAR11 Viruses in the Ocean.' *Nature* 494(7437), 357–60 (2013).

73 Mann, N., Cook, A., Millard, A. *et al.* 'Bacterial Photosynthesis Genes in a Virus.' *Nature* 424, 741 (2003).

74 See Breitbart (2012).

75 Weinmaier, T. *et al.* 'A Viability-Linked Metagenomic Analysis of Cleanroom Environments: Eukarya, Prokaryotes, and Viruses.' *Microbiome* 3 (2015). (Weinmaier and co-workers detected signatures of two phages, a Phi29-like virus and an unclassified Siphoviridae, and several viruses associated with humans and other animals, including a dragonfly virus.)

76 Yokoo, H. & Oshima, T. 'Is bacteriophage φX174 DNA a message from an extraterrestrial intelligence?' *Icarus* 38(1), 148-53 (1979).

77 Mojica, F. J. M., Juez, G., Rodriguez-Valera, F. 'Transcription at Different Salinities of Haloferax Mediterranei Sequences Adjacent to Partially Modified PstI Sites.' *Molecular Microbiology* 9, 613–21 (1993).

78 'Francis Mojica: The Modest Microbiologist.' Profile and interview with Technology Networks (2019).

79 Lander, E. S. 'The Heroes of CRISPR.' *Cell* 164(1–2), 18–28 (2016).

80 Doudna, J. *A Crack in Creation: Gene Editing and the Unthinkable Power to Control Evolution* (Houghton Mifflin, 2017).

81 The most notable abuse of CRISPR to date is the work of He Jiankui, who was sacked and sent to prison after twin girls were born from embryos he had edited with CRISPR. The procedure had been done without proper oversight or approval from the wider scientific community and the embryos did not even contain the correct HIV-resistance gene-edit he claimed to be aiming for.

82 Ireland, T. 'I Want To Help Humans Genetically Modify Themselves.' Josiah Zayner interviewed in *The Guardian*, 24 December 2017.

83 Ireland, T. Review of *Human Nature* (2015). *The Biologist* online January 2020.

84 Xiaorong, Wu *et al.* 'Bacteriophage T4 Escapes CRISPR Attack by Minihomology Recombination and Repair.' *mBio* 12(3) e0136121 (2021).

85 Landsberger, M., *et al.* 'Anti-CRISPR Phages Cooperate to Overcome CRISPR-Cas Immunity.' *Cell* 174(4), 908–16.e12 (2018).

86 Pezo, V., Jaziri, F., Bourguignon, P. Y. *et al.* 'Noncanonical DNA Polymerization by Aminoadenine-Based Siphoviruses.' *Science* 372(6541) 520–4 (2021).

87 Wein, T., & Sorek, R. 'Bacterial Origins of Human Cell-Autonomous Innate Immune Mechanisms.' *Nature Reviews Immunology* (2022).

88 Talk by Aude Bernheim at the VEGA (Viral Eco Genomics & Applications) 2021 virtual conference on 'Virus-host Molecular Warfare'. https://www.youtube.com/watch?v=T9xI6wsdgtM

89 Towse A., Hoyle C. K., Goodall J. *et al.* 'Time for a Change in How New Antibiotics Are Reimbursed: Development of an Insurance Framework for Funding New Antibiotics Based on a Policy of Risk Mitigation.' *Health Policy* 121(10), 1025–30 (2017).

90 www.citizenphage.com/

91 https://seaphages.org/

92 Little, J. S., Dedrick, R. M., Freeman, K. G. *et al.* 'Bacteriophage Treatment of Disseminated Cutaneous *Mycobacterium chelonae* Infection.' *Nature Communications* 13, 2313 (2022).

93 Turner, D., Kropinski, A. M., Adriaenssens, E. M. 'A Roadmap for Genome-Based Phage Taxonomy.' *Viruses* 13(3), 506 (2021).

94 The Actinobacteriophages Database https://phagesdb.org/phages/

95 https://twitter.com/fsantoriello/status/1466501234139938816?s=20&t=u7GZTynHoxE7EbJWfmijHw#

## Part 5

1 Pirnay, J. P. 'Phage Therapy in the Year 2035.' *Frontiers in Microbiology* 3 June 2020.

2 For various tools and platforms using very smart software to analyse, classify and make predictions about phages see iPHoP (the integrated Phage-host prediction tool); HoPhage (Host of Phage tool), MetaPhage, and Phage.ai.

3 Smith H. O., Hutchison III, C. A., Pfannkoch, C. *et al.* 'Generating a Synthetic Genome By Whole Genome Assembly: φX174 Bacteriophage from Synthetic Oligonucleotides.' *Proceedings of the National Academy of Sciences USA* 100, 15440–5 (2003).

4 Fauconnier, A., Nagel, T. E., Fauconnier, C. *et al.* 'The Unique Role That WHO Could Play in Implementing Phage Therapy to Combat the Global Antibiotic Resistance Crisis.' *Frontiers in Microbiology* 11 (2020).

5 Patz S., Becker Y., Richert-Pöggeler, K. R. *et al.* 'Phage Tail-Like Particles Are Versatile Bacterial Nanomachines – A Mini-Review.' *Journal of Advanced Research* 19, 75–84 (2019).

6 Gordillo Altamirano, F. L., Kostoulias, X., Subedi, D. *et al.* 'Phage-antibiotic Combination Is a Superior Treatment against Acinetobacter Baumannii in a Preclinical Study.' *eBioMedicine* 80, 104045 (2022).

7 Gordillo Altamirano, F. L. Forsyth, J. H., Patwa, R. *et al.* 'Bacteriophage-resistant Acinetobacter Baumannii Are Resensitized to Antimicrobials.' *Nature Microbiology* 6, 157–61 (2021), and Chan, B., *et al.* 'Phage Selection Restores Antibiotic Sensitivity in MDR *Pseudomonas aeruginosa.' Scientific Reports* 6, 26717 (2016).

8 Drexler, K. E. *Engines of Creation: The Coming Era of Nanotechnology* (Doubleday, 1986).

9 These remarkable protein pumps or 'nanomotors' are not uncommon in other forms of life – they make bacterial tails swish and move bits and bobs around in your cells too – but the ones that phages use to pack themselves up are thought to be among the

most powerful in all of nature. Pressure figure from Cruz, B., Zhu, Z., Calderer, C. *et al*. Quantitative Study of the Chiral Organization of the Phage Genome Induced by the Packaging Motor. *Biophysical Journal* 118(9), 2103–16 (2020).

10 Youle, M. *Thinking Like a Phage: The Genius of the Viruses That Infect Bacteria and Archaea* (2017).

11 Apawu, A. K., Curley, S. M., Dixon, A. R. *et al*. 'MRI Compatible MS2 Nanoparticles Designed to Cross the Blood-Brain-Barrier: Providing a Path Towards Tinnitus Treatment.' *Nanomedicine* 14(7) e-published April 2018.

12 Anand, P., O'Neil, A., Lin, E. *et al*. 'Tailored Delivery of Analgesic Ziconotide Across a Blood Brain Barrier Model Using Viral Nanocontainers.' *Scientific Reports* 5, 12497 (2015).

13 Gibb, B., Hyman, P., Schneider, C. L. 'The Many Applications of Engineered Bacteriophages – An Overview.' *Pharmaceuticals* 14, 634 (2021).

14 Bar H., Yacoby I., & Benhar, I. 'Killing Cancer Cells By Targeted Drug-Carrying Phage Nanomedicines.' *BMC Biotechnology* 8, 37 (2008).

15 Karimi, M., Mirshekari, H., Moosavi Basri, S. M. *et al*. 'Bacteriophages and Phage-Inspired Nanocarriers for Targeted Delivery of Therapeutic Cargos.' *Advance Drug Delivery Reviews* 106(Pt A), 45-62 (2016).

16 Cao, B., Li, Y., Yang, T. *et al*. 'Bacteriophage-based Biomaterials for Tissue Regeneration.' *Advanced Drug Delivery Reviews* 145, 73–95, (2019).

17 Merzlyak, A., Indrakanti, S., Lee, S. W. *et al*. 'Genetically Engineered Nanofiber-Like Viruses for Tissue Regenerating Materials.' *Nano Letters* 9(2), 846–52.

18 Liu, X., Yang, M., Lei, F. *et al*. 'Highly Effective Stroke Therapy Enabled by Genetically Engineered Viral Nanofibers.' *Advanced Materials* 34, 2201210 (2022).

19 Felix Biotech investor presentation, 2022.

20 McCarlie, S. 'A New Front'. *The Biologist* 68(3), 18–21 (2021).

21 'The PhageLand Project. The Joint Programming Initiative on

Antimicrobial Resistance' (JPIAMR). https://www.jpiamr.eu/projects/phageland/

22 Sharma, R. S., Karmakar, S., Kumar, P. *et al.* 'Application of filamentous Phages in Environment: A Tectonic Shift in the Science and Practice of Ecorestoration.' *Ecology and Evolution* 9(4), 2263–304 (2019).

23 Huss, P., & Raman, S. 'Engineered Bacteriophages as Programmable Biocontrol Agents.' *Current Opinion in Biotechnology* 61, 116–21 (2020).

24 https://www.cbd.int/climate/geoengineering/

25 Hendrix, R. W., Smith, M. C. M., Burns, R. N. *et al.* 'Evolutionary Relationships among Diverse Bacteriophages and Prophages: All the World's a Phage.' *Proceedings of the National Academy of Sciences USA* 96(5), 2192–7 (1999).

26 Paez-Espino, D., Zhou, J., Roux, S. *et al.* 'Diversity, Evolution, and Classification of Virophages Uncovered through Global Metagenomics.' *Microbiome* 7, 157 (2019).

# Index

## About the Author

Tom Ireland is a freelance science journalist and award-winning magazine editor. Tom's passion for all things microscopic began with him hiding jars of food around the house to go mouldy as a young child. From microbes to mental health, biohacking to bioethics, Tom specialises in making difficult scientific topics accessible and fun to read. As a freelance journalist he has written science stories for outlets including *BBC News*, *New Scientist* and the *Observer*. He is the editor of *The Biologist*, the magazine of the Royal Society of Biology. In 2021 he won the Giles St Aubyn Award for Non-Fiction for *The Good Virus*.